AF614769

METHODS IN MOLECULAR BIOLOGY™

Series Editor
John M. Walker
School of Life Sciences
University of Hertfordshire
Hatfield, Hertfordshire, AL10 9AB, UK

For other titles published in this series, go to
www.springer.com/series/7651

Cell-Based Assays for High-Throughput Screening

Methods and Protocols

Edited by

Paul A. Clemons*, Nicola J. Tolliday†, and Bridget K. Wagner*

**Chemical Biology Program, Broad Institute of Harvard and MIT, Cambridge, MA, USA*
†Chemical Biology Platform, Broad Institute of Harvard and MIT, Cambridge, MA, USA

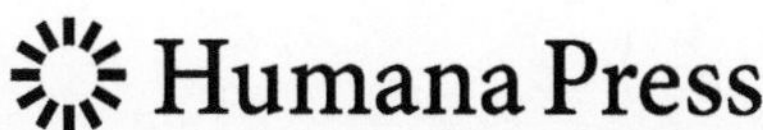

Editors
Paul A. Clemons, Ph.D.
Chemical Biology Program
Broad Institute of Harvard and MIT
Cambridge, MA
USA

Bridget K. Wagner, Ph.D.
Chemical Biology Program
Broad Institute of Harvard and MIT
Cambridge, MA
USA

Nicola Tolliday, Ph.D.
Chemical Biology Platform
Broad Institute of Harvard and MIT
Cambridge, MA
USA

ISBN: 978-1-60327-544-6 e-ISBN: 978-1-60327-545-3
ISSN: 1064-3745 e-ISSN: 1940-6029
DOI: 10.1007/978-1-60327-545-3

Library of Congress Control Number: 2008941084

Cover illustation: Chapter 4, Figure 1D

Printed on acid-free paper

springer.com

Preface

High-throughput screening (HTS) has seen over a decade of deployment in the pharmaceutical industry as an early discovery platform to feed lead compounds into pipelines culminating in the development of therapeutic agents. In recent years, interest in HTS within the academic community has increased dramatically, as an engine of discovery for potential drug candidates, for novel probes to understand fundamental biological processes, and for developing cheminformatic rules governing the interaction between chemistry and biology. A number of academic screening centers have existed for several years, including those members of the NIH-sponsored Molecular Library Probe Production Centers Network and the NCI-sponsored Initiative for Chemical Genetics, with many more facilities in the planning and start-up stages. As HTS develops into a legitimate academic pursuit, the need for reference materials regarding the philosophy and practice of screening becomes ever more pressing. Although a large proportion of screening experiments reported in the literature has been based on biochemical interactions between small molecules and purified proteins, a rich body of literature has developed around cell-based assays. We have built upon this foundation to create an easily accessible reference volume for cell-based phenotypic screening.

We encourage the reader to view this reference in a modular way. Although each chapter presents an individual protocol peculiar to the assays being discussed, in groups they represent the four governing principles of this text: (1) model biological systems, (2) screening modalities and assay systems, (3) detection technologies, and (4) approaches to data analysis. Each chapter begins with an overview of the relevant component of HTS, providing examples of its use as well as appropriate considerations and caveats. Each chapter then presents state-of-the-art methods in terms of actionable protocols; we anticipate that the reader will be interested in direct application of the methods presented.

Taken together, the methods presented in this reference can further be used in a modular fashion, culling one chapter per section to design new screens on the basis of published methods. For example, a researcher considering a fluorescent dye-based assay in mammalian cell culture might consult both Chap. 3 (Mayer et al.) on screening with mammalian cells, and Chap. 7 (An) on fluorescent dyes. Similarly, one interested in high-content imaging of zebrafish might benefit most from chapters 4 (Hong) on zebrafish screening, and 14 (Carpenter) on extracting rich information from images. We feel that this reading technique will allow researchers to take a modular approach to the design of their assays. Rather than mimic a particular screen of interest exactly, researchers might apply theory and design principles from several sections of this reference to the development of novel and creative ways of addressing biological questions using HTS.

In summary, we expect that this reference will serve three purposes. First, each chapter will present an overview of relevant approaches taken in this relatively young field. Second, each chapter will provide sufficient methodological detail to enable direct application of existing methods to new discoveries. Third, the book itself will inspire researchers to approach their screening projects in a conceptually modular fashion, enhancing the power for discovery through new combinations of existing approaches.

Cambridge, MA

Paul A. Clemons
Nicola J. Tolliday
Bridget K. Wagner

Contents

Contents

Contributors

APRIL AGEE • *Department of Botany and Plant Sciences, Imaging Center of Plant Cell Biology, University of California, Riverside, CA, USA*

W. FRANK AN • *Chemical Biology Platform, Broad Institute of Harvard and MIT, Cambridge, MA, USA*

ZOLTAN ARANY • *Cardiovascular Institute at the Beth Israel Deaconess Medical Center Dana-Farber Cancer Institute, Boston, MA, USA*

NATHALIE AULNER • *Department of Physiology and Cellular Biophysics, The Judith P. Sulzberger, M.D. Columbia Genome Center, Columbia University, New York, NY, USA*

ARUNA D. BALGI • *Department of Biochemistry and Molecular Biology, University of British Columbia, Vancouver, BC, Canada*

JAN E. BLANCHARD • *McMaster HTS Lab, McMaster University, Hamilton, ON, Canada*

ERIC D. BROWN • *Department of Biochemistry and Biomedical Sciences, McMaster HTS Lab, McMaster University, Hamilton, ON, Canada*

ANNE E. CARPENTER • *Imaging Platform, Broad Institute of Harvard and MIT, Cambridge, MA, USA*

DAVID CARTER • *Department of Botany and Plant Sciences, Imaging Center of Plant Cell Biology, University of California, Riverside, CA, USA*

EFTIHIA CAYANIS • *Department of Physiology and Cellular Biophysics, The Judith P. Sulzberger, M.D. Columbia Genome Center, Columbia University, New York, NY, USA*

GABRIELA CHIOSIS • *Program in Molecular Pharmacology and Chemistry, Memorial Sloan-Kettering Cancer Center, New York, NY, USA*

ANNIE L. CONERY • *Department of Molecular Biology, Massachusetts General Hospital, Department of Genetics, Harvard Medical School, Boston, MA, USA*

BRUCE S. EDWARDS • *Cancer Research and Treatment Center, University of New Mexico, Albuquerque, NM, USA*

CHARLES C. HONG • *Division of Cardiovascular Medicine, Department of Pharmacology, Vanderbilt University School of Medicine, Nashville, TN, USA*

IRENA IVNITSKY-STEELE • *Cancer Research and Treatment Center, University of New Mexico, Albuquerque, NM, USA*

SERENE JOSIAH • *Cambridge, MA, USA*

ADAM B. KEETON • *Southern Research Molecular Libraries Screening Center, Assay Implementation Group, Southern Research Institute, Birmingham, AL, USA*

ANDY M.F. LIU • *Department of Biochemistry, Molecular Neuroscience Center, Biotechnology Research Institute, Hong Kong University of Science and Technology, Hong Kong, China*

RICO K.H. LO • *Department of Biochemistry, Molecular Neuroscience Center, Biotechnology Research Institute, Hong Kong University of Science and Technology, Hong Kong, China*

THOMAS MAYER • *Department of Physiology and Cellular Biophysics, The Judith P. Sulzberger, M.D. Columbia Genome Center, Columbia University, New York, NY, USA*

TERENCE I. MOY • *Department of Molecular Biology, Massachusetts General Hospital, Department of Genetics, Harvard Medical School, Boston, MA, USA*

DAVID C. NEW • *Department of Biochemistry, Molecular Neuroscience Center, Biotechnology Research Institute, Hong Kong University of Science and Technology, Hong Kong, China*

EYLEEN O' ROURKE • *Department of Molecular Biology, Massachusetts General Hospital, Department of Genetics, Harvard Medical School, Boston, MA, USA*

ERIC R. PROSSNITZ • *Cancer Research and Treatment Center, University of New Mexico, Albuquerque, NM, USA*

MICHEL ROBERGE • *Department of Biochemistry and Molecular Biology, University of British Columbia, Vancouver, BC, Canada*

LARRY A. SKLAR • *Cancer Research and Treatment Center, University of New Mexico, Albuquerque, NM, USA*

DEBORAH H. SMITH • *Department of Physiology and Cellular Biophysics, The Judith P. Sulzberger, M.D. Columbia Genome Center, Columbia University, New York, NY, USA*

UDO TOEBBEN • *Department of Physiology and Cellular Biophysics, The Judith P. Sulzberger, M.D. Columbia Genome Center, Columbia University, New York, NY, USA*

NICOLA J. TOLLIDAY • *Chemical Biology Platform, Broad Institute of Harvard and MIT, Cambridge, MA, USA*

BRIDGET K. WAGNER • *Chemical Biology Program, Broad Institute of Harvard and MIT, Cambridge, MA, USA*

YUNG H. WONG • *Department of Biochemistry, Molecular Neuroscience Center, Biotechnology Research Institute, Hong Kong University of Science and Technology, Hong Kong, China*

MICHAEL R. WYLER • *Department of Physiology and Cellular Biophysics, The Judith P. Sulzberger, M.D. Columbia Genome Center, Columbia University, New York, NY, USA*

RICHARD D. YE • *Cancer Research and Treatment Center, University of New Mexico, Albuquerque, NM, USA*

SUSAN M. YOUNG • *Cancer Research and Treatment Center, University of New Mexico, Albuquerque, NM, USA*

SOUMAYA ZLITNI • *Department of Biochemistry and Biomedical Sciences, McMaster University, Hamilton, ON, Canada*

Chapter 1

Introduction: Cell-Based Assays for High-Throughput Screening

W. Frank An and Nicola J. Tolliday

Summary

Cell-based assays represent approximately half of all high-throughput screens (HTS) currently performed. Here we review the history and status of HTS, and summarize some of the challenges and benefits associated with the use of cell-based assays in HTS, drawing upon themes that will reemerge in subsequent chapters in this book. Approaches for successful experimental design and execution of cell-based HTS are introduced, including strategies for assay development, implementation of primary and secondary screens, and target identification. In doing so, we hope to provide a comprehensive review of the cell-based HTS process and an introduction to the methodologies and techniques described in this book.

Key words: Assay development, Cell-based assay, High-content screening, High-throughput screening, Small-molecule screening, Target identification.

1. High-Throughput Screening: A Brief History

High-throughput screening (HTS) typically refers to a process in which large numbers of chemicals are tested (i.e., screened) with high efficiency to identify biologically active small molecules as candidates for further validation in additional biological or pharmacological experiments. In the pharmaceutical and biotechnology industries, this process normally involves large compound collections (hundreds of thousands to millions of small molecules), industrial-scale robots, and sophisticated automation. Ultimately, the goal of HTS is to generate chemical structures (leads) that will feed into the drug discovery and development pipeline in a therapeutics setting, or that will be used as probes to address biological questions in basic research (tool compounds).

P.A. Clemons et al. (eds.), *Cell-Based Assays for High-Throughput Screening, Methods in Molecular Biology, Vol. 486*

DOI: 10.1007/978-1-60327-545-3_1

HTS has seen its emergence and maturation over the past two decades as an early discovery platform in the pharmaceutical industry. In the late 1980s to mid-1990s, significant advances in combinatorial chemistry and genomics helped drive a rapid growth in HTS. The potential to deliver thousands of novel compounds quickly and cheaply generated much optimism about the promise of future drug discovery and helped drive development of HTS technologies to evaluate the large numbers of new compounds available through combinatorial chemistry. Around the same time, rapid progress in genomics revealed many potential new drug targets. A lack of documented "druggability" and structural information for many such novel genomic targets *(1)* led to HTS becoming the method of choice for identification of small-molecule modulators of these targets from within ever-increasing compound collections in the pharmaceutical industry. As a result, the pharmaceutical industry invested large sums of capital to fuel rapid advances in HTS technologies in terms of automation, miniaturization, and assay methodology *(2, 3)*. Efficiency and throughput have been greatly improved. The 384-well plate format is now standard for HTS, with an increasing number of screens using a 1,536-well plate format or even higher well-density format (ultra-HTS or uHTS). Throughputs of ≥100,000 compounds screened per day are routine in lead HTS practitioner laboratories, where up to 50 HTS campaigns are being run each year *(2, 4)*. Return on these investments is evidenced by the increasing numbers of leads, clinical candidates, and marketed drugs arising from HTS. For example, Virumune, an anti-HIV drug, was a direct result of early HTS efforts *(5, 6)*. Additionally, advances in HTS technologies have been of benefit to other fields. Approaches developed for small-molecule screening have also been applied to identification and validation of specific gene functions among collections of cDNAs *(7, 8)* and RNA interference (RNAi) reagents *(9)*.

In recent years, interest in HTS among academic researchers has increased dramatically. There are now in existence many academic screening centers, to which many of the principles and lessons learned within the drug-discovery community are applied (a comprehensive database of academic screening centers and their capabilities is hosted by the Society for Biomolecular Sciences, http://www.sbsonline.org/). Impressive progress has been made in terms of automation, throughput, and improved screen paradigms *(10)*. In addition to providing HTS resources to the academic community for identification of both probes and leads for drug discovery, many of these centers promote open-source data sharing of small-molecule screening data. For example, *ChemBank* (http://chembank.broad.harvard.edu), from the National Cancer Institute's Initiative for Chemical Genetics *(11)*, and PubChem (http://pubchem.ncbi.nlm.nih.gov), from the National Institutes of Health's Molecular Libraries Roadmap, make freely available data regarding small molecules and small-molecule screens, with the aim of facilitating both basic and applied research in the global scientific community.

2. Assay Types: Cell-Based vs. Biochemical Screens

Assays developed for HTS can be divided very broadly into two categories: biochemical assays and cell-based assays. Biochemical assays are target based and historically have been the mainstay of HTS campaigns in the pharmaceutical industry. Such in vitro assays include assessment of enzymatic activity (e.g., for kinases *(12)*, proteases *(13)*, or transferases *(14)*), receptor–ligand binding (e.g., for G-protein coupled receptors (GPCRs) *(15)*, ion channels *(16)*, or nuclear receptors *(17)*), or protein–protein interactions *(18)*. Biochemical assays are often direct and specific to the target of interest and can be miniaturized readily with less variability owing to the homogeneous nature of reactions. However, not all targets can be purified or prepared in a manner suitable for biochemical measurement. Additionally, activity of a small molecule in a reconstituted in vitro assay does not always translate into the same activity in a cellular context, because of issues including membrane permeability, off-target effects, and cytotoxicity.

In recent years, cell-based assays have emerged as a more physiological alternative to assays involving purified proteins. In contrast to biochemical target-based assays, cell-based assays often assume no a priori knowledge of a *direct* molecular target. Instead, many cell-based assays aim to identify modulators of a *pathway* of interest in the more physiological environment of a cell, complete with intact regulatory networks and feedback control mechanisms *(19)*. Examples of cell-based assays include functional assays (e.g., second messenger mobilization after GPCR activation *(20, 21)*), reporter gene assays *(22, 23)*, and phenotypic assays for cellular processes (e.g., cell migration *(24)* or cytokinesis *(25)*). In such assays, entire pathways of interest can be interrogated, providing the opportunity for multiple potential intervention points, as opposed to a single predefined step with the biochemical approach. This approach not only expands the repertoire of targets, but also provides additional chemical structures as plausible starting points for lead identification (for example, *see (26)*). Moreover, cell-based assays allow for the selection of compounds that can cross cellular membranes and can also provide indications of acute cytotoxicity as an early alert for later-stage lead/probe-discovery efforts. Today, most currently available instrumentation in tissue culture, including automated cell culture (such as Select™, The Automation Partnership) and liquid handling, in conjunction with careful assay development, allows execution of high-quality cell-based screens. Cell-based HTS in 384, 1,536, and even 3,456-well plate format has been reported *(27–29)*.

Whole organism-based screens provide an additional level of physiological relevance beyond cell-based assays. Unicellular model organisms (e.g., yeast, bacteria) are readily adapted to grow in microtiter plates; Brown and colleagues and Balgi and Roberge provide comprehensive examples of microbial assays in Chaps. 2 and 9, respectively. Multicellular organisms, with intact cell-to-cell communication and three-dimensional tissue organization, represent additional challenges for adaptation to HTS. However, several model metazoan systems have been successfully adapted for use in small-molecule screens. Examples using zebrafish, worms, and plants in HTS are provided in Chaps. 4–6.

3. Experimental Design and Planning

Several important factors need to be considered when planning a small-molecule screen. The first step is to identify a screening strategy that maintains the appropriate biological context while balancing feasibility in terms of reagent availability and adaptation to automation. For cell-based assays, these considerations includes choice of biological system (primary cell, native or engineered cell line, or model organism), choice of assay approach (functional, reporter-gene, or phenotypic), and assay readout (uniform well readout or high content). Further considerations include follow-up biological experiments (counter screens, secondary assays, target identification, and in vivo validation) and determination of data-analytical strategies for interpretation of data arising from assay development, HTS, and follow-up assays (reviewed in more detail in Chap. 14). In Chap. 3, Mayer and colleagues provide an example of the use of primary mammalian cells for HTS, in the context of the ERK signaling pathway, and use of an engineered cell line with a fluorescent image-based readout is described by An in Chap. 7.

A variety of detection methods can be used for cell-based assays. These break down broadly into uniform well measurements (one measurement per well, representing a population average) and high-content measurements (multiple measurements per well, often representing subpopulations of cells, or even subcellular features). Uniform well measurements can be obtained using fluorescence, luminescence, and spectrophotometric methods. Among them, fluorescent detection is arguably the most widely used approach owing to its high sensitivity, the diverse selection of fluorophores available, ease of operation, and various readout modes, such as fluorescence intensity, fluorescence polarization, fluorescence resonance energy transfer (FRET), fluorescence lifetime, time-resolved fluorescence, and combinations of these techniques, such as time-resolved fluorescence polarization. In Chap. 7, An provides two experimental protocols for fluorescence-based assays.

High-content screening (HCS) refers to any technique or process in which multiple measurements are obtained from a single well. Compared with single-parameter assays, HCS provides richer contextual and concurrent information that helps better illustrate both the behavior and the mechanism of action of small molecules and genetic manipulations. The most popular HCS format is image-based screening using an automated microscope (for examples of HCS imaging methods, *see (30)*), and to some researchers HCS is synonymous with image-based screening. In this textbook, however, the broader definition of HCS is used. In Chap. 12, Edwards and colleagues describe an application of high-throughput flow cytometry to identify selective ligands for two related G-protein coupled receptors.

4. Assay Development and Readiness for HTS

Assay development refers to the process in which potential approaches for measuring a particular target or biological process are evaluated; the best approach is further optimized in terms of throughput, cost, sensitivity, and signal dynamic range and variation, and adapted to the instrumentation of the screening facility. Assay development is critical for successful HTS, not only as applied to the primary screen but also for medium-throughput follow-up assays (secondary assays, selectivity assays, and other compound-profiling assays) that use HTS instrumentation to analyze hundreds to a few thousands of compounds. Given the costly nature of HTS campaigns (estimates for fully-loaded screening costs range from $0.50 to $1.00 per well, or $250,000–$500,000 for a 500,000 compound screen), it is worthwhile to spend time in advance to ensure that high-quality data of biological relevance are produced from HTS.

Because HTS tests hundreds of thousands to millions of compounds at once, it often requires additional considerations than development of a bench-top assay. One of the first tasks is to identify optimal conditions using the prototype assay. A full range of experimental conditions needs to be evaluated to determine the best signal-to-background levels. For cell-based assays, optimization includes, but is not limited to, titration of cell density, titration of assay reagent(s), determination of optimal concentrations of modulator(s) (for small-molecule modifier screens *(31)*), and determination of incubation time with compounds. Positive controls are best included and titrated to provide diagnostic information for each experiment. Additionally, the stability of both assay reagents and readout signals should be evaluated over the time course of an assay with the eventual HTS process in mind. For example, time lapse between the first and last well on a plate, as well as the first and last plate of a batch, needs to be taken into consideration when assessing reagent and signal stability. Sensitivity to compound solvent (usually dimethylsulfoxide (DMSO)) should also be determined for cell-based assays. We recommend that a titration of up to 1% DMSO be performed.

Another early task is miniaturization. If we use 384-well plates as an example, typical reaction volumes are 25–50 μL per well. The goal at this stage is to maintain robust signal detection and acceptable signal-to-background ratios despite considerable reduction of reaction volume, reagent amounts, and cell numbers. There is no absolutely correct order of condition optimization and miniaturization, as long as the final assay is robust in 384-well plate format. In general, however, proceeding in ascending order of parameter contribution to assay noise will make assay development results easier to interpret. Yet another goal is to assess the liquid-handling compatibility with the assay. While most current liquid-handling equipment interfaces well with the 384-well plate format, accuracy and precision when dispensing low volumes of liquid (1–5 μL) can be a concern and should be tested. Instrument settings, such as volume, speed, and height and position of the pins/tips, need to be adjusted to achieve proper distribution and aspiration of the reagents/solutions while avoiding splashing or damaging monolayers of adherent cells by mechanical force.

Reproducibility is another critical factor. Well-to-well, plate-to-plate, day-to-day, and batch-to-batch (protein or cells) variations should be tested using positive (when a positive control is available) and DMSO-only controls. At the Broad Institute HTS facility, we recommend using a 384-well compound plate comprising half positive control (e.g., columns 1–12) wells and half DMSO-containing wells (e.g., columns 13–24) to evaluate reproducibility. This compound plate can then be pin-transferred into replicate assay plates on two consecutive days to evaluate well-to-well, plate-to-plate, and day-to-day variations of the assay. Alternatively, dose responses of positive controls can be evaluated on a day-to-day and batch-to-batch basis.

The most widely accepted measurement of assay quality and readiness is the Z' factor *(32)*. This metric quantifies the separation of a positive activity (sample) and background control in the absence of intervention of test compounds. It is determined as follows:

$$Z' = 1 - \frac{3 \times (\sigma_s + \sigma_c)}{|\mu_s - \mu_c|}$$

where σ_s and σ_c are the standard deviations of the sample and the control, respectively, and μ_s and μ_c are the means of sample and control, respectively. $Z' \geq 0.5$ indicates an excellent assay. An assay with $0 < Z' < 0.5$ is considered marginal; it may be suitable for HTS, but further optimization is often required. Assays with $Z' \leq 0$ are not suitable for HTS. Because Z' is based on standard deviations, large numbers of replicates are needed to compute meaningful Z' values. At the Broad Institute HTS facility, a minimum of 96 wells of a 384-well plate are used for Z' calculations. After the assay has achieved an acceptable Z' value, a pilot screen of approximately 2,000–10,000 compounds can be run to validate the high-throughput assay.

5. Execution of HTS and Beyond

The execution phase of HTS also requires careful planning. Screeners need to ensure that they have a sufficient and timely supply of target materials (proteins, cells, membrane preparations), consumables (plates, tips), and reagents, not only for the primary screen but also for retesting, EC_{50} determinations. and secondary assays. At the Broad Institute, primary screens are typically performed in duplicate, and IC_{50} or EC_{50} information is obtained using a duplicate, eight-point dose–response scheme at the retesting stage. Analysis, interpretation, and mining of HTS data represent other critical components for successful screens. Appropriate strategies for data analysis and interpretation should be considered in advance for all stages of the HTS process. In Chap. 14, Josiah discusses parameters used to design plate maps and controls and describes how to use the information obtained from those controls to assess screen performance. Specific image-based HCS analysis considerations are highlighted in Chaps. 11 and 15.

6. Target Identification

Once interesting “hits” are identified in a phenotypic cell-based assay, one frequently asked question is what are the protein targets? For cell-based assays in which a target of interest is overexpressed, e.g., receptor function assays, the attribution of “hit” activity to the overexpressed target is easier to ascertain, and the remaining question is whether there are any off-target events that lead to the same phenotype. On the other hand, cell-based assays that screen for a phenotype (e.g., cell cycle arrest or reporter activity) instead of a target mean that many intervention points or protein candidates could conceivably produce the same phenotype. In this latter scenario, target identification becomes a more pressing issue in order to understand the mechanism of action of the "hits," and to facilitate the downstream discovery process.

There have been many successful examples of identification of the protein target(s) of bioactive small molecules. Among the most ground breaking are the discovery of (1) immunophilins as targets of a group of naturally occurring immunosuppressants: cyclophilin A as the target of cyclosporine A (CsA) *(33–36)*, and FKBP12 as the target of FK506 and rapamycin *(37–39)*; and (2) the cloning and identification of the first histone deacetylase (HDAC) as the target of trapoxin, a small molecule that increases cellular histone acetylation levels and causes cell cycle arrest *(40)*. The immunophilin discovery provided profound insight into mechanisms of action of widely used clinical immunosuppressants and the signaling cascade in T-cell signal transduction, while the isolation and cloning of the first HDAC catalyzed much of the subsequent research into chromatin and its function as a key regulatory element.

While there have been many examples of ad hoc successes in target identification for bioactive small molecules, the field is still struggling to find a systematic, case-independent, broadly applicable approach with a high success rate *(41)*. One of the major challenges is to devise methods through which investigators can reliably identify not only highly abundant proteins as targets for high-affinity bioactive small molecules, which happened to be the case for many of the historic successes, but also lower-abundance proteins for medium-affinity bioactive small molecules. While there may not be a panacea in target identification for all bioactive small molecules, a number of recent studies show progress towards a systematic approach, as described below.

The affinity-based approach is the most direct way to identify proteins that bind bioactive small molecules. This approach usually involves chemical modification of the small molecule of interest (bait compound) for attachment through a linker to a solid support (for pull-down-type experiments) or by introducing to it a reactive group that can be activated to bind its protein target (e.g., aryl azide for photo-affinity labeling *(42)*). Usually the bait compound also has a detection tag such as ^{125}I, ^{3}H, or biotin, or functionality that can react specifically with other chemical groups on detection agents, e.g., terminal acetylenes that react with azide-modified rhodamine by the copper(I)-catalyzed 1,2,3-triazole formation click chemistry *(43, 44)*. One common requirement of these methods is the understanding of the structure–activity relationship (SAR) of the bioactive small molecule, such that modification does not interfere with the known activity of the bioactive small molecule. New ways to create high-affinity probes from low-affinity small-molecule probes without tedious chemistry–biology iterations have also been described which involve fusion of two modest small-molecule probes *(45–47)*.

A promising proteomics-based technique for detecting specific bioactive small molecule-interacting proteins via the affinity approach is stable isotope labeling by amino acids in cell culture (SILAC) *(41, 48)*. SILAC starts by labeling proteins from cells grown separately in heavy amino acid-supplemented medium and in light (normal) amino acid-supplemented medium. The heavy protein preparation and the light protein preparation are made from these otherwise identical culture conditions and then subject to chromatography through a bioactive small-molecule affinity matrix and a control compound affinity matrix, respectively. Eluents of both are then mixed and examined by mass spectrometry for ratio of abundance. For proteins specifically retained by the bioactive small-molecule matrix, there should be more heavy signal relative to light signal, whereas for nonspecific binders to the matrix the heavy:light ratio should be close to unity. Better sensitivity and higher throughput are two advantages of this ratiometric, mass spectrometry-based approach as opposed to traditional gel-based approaches.

Other complementary, systematic schemes to the affinity-based approach in target identification include genetics and function/phenotype association investigations. Genome-wide heterozygote hypersensitivity to small molecules in yeast has been used to suggest drug targets *(49–51)*, and the same principle has been applied to mammalian cells in gene-dosage screens *(52)*. A yeast three-hybrid system has also been reported, where the reporter activity is controlled by potential protein targets that interact with the bioactive small molecule in question *(53)*. The advent of genome-wide knockdown by RNAi should facilitate these systematic genetic approaches and provide a wealth of phenotypes that can be compared with those achieved by bioactive small molecules. With increased realization in the academic and pharmaceutical community that target identification represents a critical challenge, and with the increased activities to address the issues associated, it is likely that additional innovative, systematic approaches will be reported in the near future.

7. Summary

HTS has matured over the past two decades into an indispensable part of drug discovery and basic research in both the pharmaceutical industry and academia. Cell-based assays, a focus of this textbook with comprehensive chapters describing topics from screening methodology to data analysis, account for approximately half of all HTS campaigns and present exciting potential for breakthroughs in both our understanding of biology and expansion of our arsenal of small-molecule tools and drugs. Granted, there are challenges, such as systematic target identification following certain cell-based screens. However, we hope that the detailed techniques and methodologies provided in this book will enable researchers to practice the protocols in a HTS setting, and that such practice will lead to new innovations that fulfill the promise of cell-based assays.

Acknowledgements

We thank Drs M. Schenone and I. Smukste for insightful discussions on target identification. The work has been funded in whole or in part with federal funds from the National Cancer Institute's Initiative for Chemical Genetics, National Institutes of Health, under Contract No. N01-CO-12400. The content of this publication does not necessarily reflect the views or policies of the Department of Health and Human Service, nor does mention of trade names, commercial products, or organizations imply endorsement by the U.S. Government.

References

1. Hopkins, A. L. and Groom, C. R. (2002) The druggable genome. *Nat. Rev. Drug Discov.* 1, 727–30.
2. Beggs, M. (2000) HTS-where next. *Drug Discov. World* winter, 25–30.
3. Hertzberg, R. P. and Pope, A. J. (2000) High-throughput screening: New technology for the 21st century. *Curr. Opin. Chem. Biol.* 4, 445–451.
4. Liu, B., Li, S., and Hu, J. (2004) Technological advances in high-throughput screening. *Am. J. Pharmacogenomics* 4, 263–276.
5. Grozinger, K., Proudfoot, J., and Hargrave, K. (2006) Discovery and development of nevirapine. In Chorghade, M. S. (ed.) Drug Discovery and Development. Wiley-VCH, Weinheim, Germany, pp. 353–363.
6. Kell, D. (1999) Screensavers: Trends in high-throughput analysis. *Trends Biotech.* 17, 89.
7. Zitzler, J., Link, D., Schafer, R., Liebetrau, W., Kazinski, M., Bonin-Debs, A., et al. (2004) High-throughput functional genomics identifies genes that ameliorate toxicity due to oxidative stress in neuronal HT-22 cells: GFPT2 protects cells against peroxide. *Mol. Cell. Proteomics* 3, 834–840.
8. Korherr, C., Gille, H., Schafer, R., Koenig-Hoffmann, K., Dixelius, J., Egland, K. A., et al. (2006) Identification of proangiogenic genes and pathways by high-throughput functional genomics: TBK1 and the IRF3 pathway. *Proc. Natl. Acad. Sci. U.S.A.* 103, 4240–4245.
9. Moffat, J., Grueneberg, D. A., Yang, X., Kim, S. Y., Kloepfer, A. M., Hinkle, G., et al. (2006) A lentiviral RNAi library for human and mouse genes applied to an arrayed viral high-content screen. *Cell* 124, 1283–1298.
10. Inglese, J., Auld, D. S., Jadhav, A., Johnson, R. L., Simeonov, A., Yasgar, A. et al. (2006) Quantitative high-throughput screening: A titration-based approach that efficiently identifies biological activities in large chemical libraries. *Proc. Natl. Acad. Sci. U.S.A.* 103, 11473–11478.

11. Tolliday, N., Clemons, P. A., Ferraiolo, P., Koehler, A. N., Lewis, T. A., Schreiber, S. L., et al. (2006) Small molecules, big players: The national cancer institute';s initiative for chemical genetics. *Cancer Res.* 66, 8935–8942.
12. Burns, S., Travers, J., Collins, I., Rowlands, M. G., Newbatt, Y., Thompson, N., et al. (2006) Identification of small-molecule inhibitors of protein kinase B (PKB/AKT) in an alphascreenTM high-throughput screen. *J. Biomol. Screen.* 11, 822–827.
13. Sudo, K., Yamaji, K., Kawamura, K., Nishijima, T., Kojima, N., Aibe, K., et al. (2005) High-throughput screening of low molecular weight NS3-NS4A protease inhibitors using a fluorescence resonance energy transfer substrate. *Antivir. Chem. Chemother.* 16, 385–392.
14. Swaney, S., McCroskey, M., Shinabarger, D., Wang, Z., Turner, B. A., and Parker, C. N. (2006) Characterization of a high-throughput screening assay for inhibitors of elongation factor P and ribosomal peptidyl transferase activity. *J. Biomol. Screen.* 11, 736–742.
15. Allen, M., Reeves, J., and Mellor, G. (2000) High throughput fluorescence polarization: A homogeneous alternative to radioligand binding for cell surface receptors. *J. Biomol. Screen.* 5, 63–69.
16. Xu, J., Wang, X., Ensign, B., Li, M., Wu, L., Guia, A., et al. (2001) Ion-channel assay technologies: Quo vadis? *Drug Discov. Today* 6, 1278–1287.
17. Parker, G. J., Law, T. L., Lenoch, F. J., and Bolger, R. E. (2000) Development of high throughput screening assays using fluorescence polarization: nuclear receptor-ligand-binding and kinase/phosphatase assays. *J. Biomol. Screen.* 5, 77–88.
18. Kenny, C. H., Ding, W., Kelleher, K., Benard, S., Dushin, E. G., Sutherland, A. G., et al. (2003) Development of a fluorescence polarization assay to screen for inhibitors of the FtsZ/ZipA interaction. *Anal. Biochem.* 323, 224–233.
19. Clemons, P. A. (2004) Complex phenotypic assays in high-throughput screening. *Curr. Opin. Chem. Biol.* 8, 334–338.
20. Chambers, C., Smith, F., Williams, C., Marcos, S., Zhen, S., Hayter, P., et al. (2003) Measuring intracellular calcium fluxes in high throughput mode. *Comb. Chem. High Throughput Screen* 6, 355–362.
21. Kariv, I., Stevens, M. E., Behrens, D. L., and Oldenburg, K. R. (1999) High throughput quantitation of cAMP production mediated by activation of seven transmembrane domain receptors. *J. Biomol. Screen.* 4, 27–32.
22. Li, X., Shen, F., Zhang, Y., Zhu, J., Huang, L., and Shi, Q. (2007) Functional characterization of cell lines for high-throughput screening of human neuromedin U receptor subtype 2 specific agonists using a luciferase reporter gene assay. *Eur. J. Pharm. Biopharm.* 67, 284–292.
23. Beck, V., Pfitscher, A., and Jungbauer, A. (2005) GFP-reporter for a high throughput assay to monitor estrogenic compounds. *J. Biochem. Biophys. Methods* 64, 19–37.
24. Yarrow, J. C., Totsukawa, G., Charras, G. T., and Mitchison, T. J. (2005) Screening for cell migration inhibitors via automated microscopy reveals a Rho-kinase inhibitor. *Chem. Biol.* 12, 385–395.
25. Eggert, U. S., Kiger, A. A., Richter, C., Perlman, Z. E., Perrimon, N., Mitchison, T. J., et al. (2004) Parallel chemical genetic and genome-wide RNAi screens identify cytokinesis inhibitors and targets. *PLoS Biol.* 2, 2135–1243.
26. Krejci, P., Pejchalova, K., and Wilcox, W. R. (2007) Simple, mammalian cell-based assay for identification of inhibitors of the Erk MAP kinase pathway. *Invest. New Drugs* 25, 391–394.
27. Bradley, J., Gill, J., Bertelli, F., Letafat, S., Corbau, R., Hayter, P., et al. (2004) Development and automation of a 384-well cell fusion assay to identify inhibitors of CCR5/CD4-mediated HIV virus entry. *J. Biomol. Screen.* 9, 516–524.
28. Wunder, F., Stasch, J. P., Hutter, J., Alonso-Alija, C., Huser, J., and Lohrmann, E. (2005) A cell-based cGMP assay useful for ultra-high-throughput screening and identification of modulators of the nitric oxide/cGMP pathway. *Anal. Biochem.* 339, 104–112.
29. Brandish, P. E., Chiu, C. S., Schneeweis, J., Brandon, N. J., Leech, C. L., Kornienko, O., et al. (2006) A cell-based ultra-high-throughput screening assay for identifying inhibitors of D-amino acid oxidase. *J. Biomol. Screen.* 11, 481–487.
31. Koeller, K. M., Haggarty, S. J., Perkins, B. D., Leykin, I., Wong, J.-C., Kao, M. C. J., et al. (2003). Chemical genetic modifier screens: Small molecule trichostatin suppressors as probes of acetylation in transcription, cell cycle progression, and stability of the cytoskeleton. *Chem. Biol.* 10, 397–410.
32. Zhang, J.-H., Chung, T. D. Y., and Oldenburg, K. R. (1999) A simple statistical parameter for use in evaluation and validation of high throughput screening assays. *J. Biomol. Screen.* 4, 67–73.

33. Harding, M. W., Handschumacher, R. E., and Speicher, D. W. (1986) Isolation and amino acid sequence of cyclophilin. *J. Biol. Chem.* 261, 8547–8555.
34. Handschumacher, R. E., Harding, M. W., Rice, J., Drugge, R. J., and Speicher, D. W. (1984) Cyclophilin: A specific cytosolic binding protein for cyclosporin A. *Science* 226, 544–547.
35. Fischer, G., Wittmann-Liebold, B., Lang, K., Kiefhaber, T., and Schmid, F. X. (1989) Cyclophilin and peptidyl-prolyl cis-trans isomerase are probably identical proteins. *Nature* 337, 476–478.
36. Takahashi, N., Hayano, T., and Suzuki, M. (1989) Peptidyl-prolyl cis-trans isomerase is the cyclosporin A-binding protein cyclophilin. *Nature* 337, 473–475.
37. Lane, W. S., Galat, A., Harding, M. W., and Schreiber, S. L. (1991) Complete amino acid sequence of the FK506 and rapamycin binding protein, FKBP, isolated from calf thymus. *J. Protein. Chem.* 10, 151–160.
38. Harding, M. W., Galat, A., Uehling, D. E., and Schreiber, S. L. (1989) A receptor for the immunosuppressant FK506 is a cis-trans peptidyl-prolyl isomerase. *Nature* 341, 758–760.
39. Siekierka, J. J., Hung, S. H. Y., Poe, M., Lin, C. S., and Sigal, N. H. (1989) A cytosolic binding protein for the immunosuppressant FK506 has peptidyl-prolyl isomerase activity but is distinct from cyclophilin. *Nature* 341, 755–757.
40. Taunton, J., Hassig, C. A., and Schreiber, S. L. (1996) A mammalian histone deacetylase related to the yeast transcriptional regulator Rpd3p. *Science* 272, 408–411.
41. Burdine, L. and Kodadek, T. (2004) Target identification in chemical genetics: The (often) missing link. *Chem. Biol.* 11, 593–597.
42. Colca, J. R. and Harrigan, G. G. (2004) Photo-affinity labeling strategies in identifying the protein ligands of bioactive small molecules: Examples of targeted synthesis of drug analog photoprobes. *Comb. Chem. High Throughput Screen.* 7, 699–704.
43. Kolb, H. C. and Sharpless, K. B. (2003) The growing impact of click chemistry on drug discovery. *Drug Discov. Today* 8, 1128–1137.
44. Speers, A. E. and Cravatt, B. F. (2004) Profiling enzyme activities *in vivo* using click chemistry methods. *Chem. Biol.* 11, 535–546.
45. Maly, D. J., Choong, I. C., and Ellman, J. A. (2000) Combinatorial target-guided ligand assembly: Identification of potent subtype-selective c-Src inhibitors. *Proc. Natl. Acad. Sci. U.S.A.* 97, 2419–2424.
46. Sem, D. S., Bertolaet, B., Baker, B., Chang, E., Costache, A. D., Coutts, S., et al. (2004) Systems-based design of bi-ligand inhibitors of oxidoreductases: Filling the chemical proteomic toolbox. *Chem. Biol.* 11, 185–194.
47. Profit, A. A., Lee, T. R., and Lawrence, D. S. (1999) Bivalent inhibitors of protein tyrosine kinases. *J. Am. Chem. Soc.* 121, 280–283.
48. Ong, S. E., Blagoev, B., Kratchmarova, I., Kristensen, D. B., Steen, H., Pandey, A., et al. (2002) Stable isotope labeling by amino acids in cell culture, SILAC, as a simple and accurate approach to expression proteomics. *Mol. Cell. Proteomics* 1, 376–386.
49. Lum, P. Y., Armour, C. D., Stepaniants, S. B., Cavet, G., Wolf, M. K., Butler, J. S., et al. (2004) Discovering modes of action for therapeutic compounds using a genome-wide screen of yeast heterozygotes. *Cell* 116, 121–137.
50. Giaever, G., Flaherty, P., Kumm, J., Proctor, M., Nislow, C., Jarmaillo, D. F., et al. (2004) Chemogenomic profiling: Identifying the functional interactions of small molecules in yeast. *Proc. Natl. Acad. Sci. U.S.A.* 101, 793–798.
51. Perlstein, E. O., Ruderfer, D. M., Roberts, D. C., Schreiber, S. L., Kruglyak, L. (2007) Genetic basis of individual differences in the response to small-molecule drugs in yeast. *Nat. Genet.* 39, 496–502.
52. Luesch, H. (2006) Towards high-throughput characterization of small molecule mechanisms of action. *Mol. BioSyst.* 2, 609–620.
53. Kley, N. (2004) Chemical dimerizers and three-hybrid systems: Scanning the proteome for targets of organic small molecules. *Chem. Biol.* 11, 599.

Chapter 2

High-Throughput Screening of Model Bacteria

Soumaya Zlitni, Jan E. Blanchard, and Eric D. Brown

Summary

Small-molecule screening campaigns of model bacteria have been conducted extensively in biotechnology and pharmaceutical companies to search for novel compounds with antibacterial activity. Recently, there has been increasing interest in running such high-throughput screens within academic settings to answer questions in biology. In this respect, whole-cell screening has the particular advantage of identifying compounds with physical and chemical properties compatible with microbial cell permeation, thereby providing probes with which to study diverse aspects of microbial cell physiology and biochemistry. The focus of this chapter is to describe a general method of running a high-throughput screen against a model bacterium to identify small molecules with growth inhibitory activity. Once the primary bioactives have been identified, the determination of their dose–response relationships with the target microbe further characterizes their growth inhibitory effect.

Key words: Small-molecule screening, Model bacteria, Small-molecule library, Growth inhibition, Primary screening, Minimum inhibitory concentration, EC_{50}, Median effective concentration, *E. coli*.

1. Introduction

1.1. Small-Molecule Screens of Model Bacterial Systems

We restrict our scope in this chapter to small-molecule screening of model bacteria, e.g., *Escherichia coli* and *Bacillus subtilis.* The study of these model microbes has contributed greatly to fundamental knowledge in cell biology as well as an understanding of bacterial physiology, infectious disease, immunology, and genetic engineering. Owing primarily to ease of culture and genetic tractability, model microbes have attracted enormous experimental attention and have provided extraordinary insights into basic biology.

Among the prokaryotes, *E. coli* and *B. subtilis* represent highly tractable models for Gram-negative and Gram-positive bacteria,

P.A. Clemons et al. (eds.), *Cell-Based Assays for High-Throughput Screening, Methods in Molecular Biology, Vol. 486*

DOI: 10.1007/978-1-60327-545-3_1

respectively. While the study of *E. coli* has fundamentally transformed the field of molecular biology and genetic engineering, it has likewise provided extraordinary advances in our understanding of basic bacterial physiology. *E. coli* has also proven to be an ideal model for the study of processes fundamental to all life, including DNA replication, protein translation, and general metabolism. *E. coli* was among the first organisms for which genome sequence information became available. In addition to classical genetics, state-of-the-art tools for its study continue to emerge. These include highly annotated sequence information *(1)* as well as genome-scale gene-deletion *(2)* and protein-expression libraries *(3)*. These new tools will become increasingly important in forward chemical-genetic studies to sort out mechanisms of action of biologically active molecules.

While screening in bacterial systems has most often been implemented to find small molecules that inhibit growth, such efforts have been limited for the most part to researchers in pharmaceutical and biotechnology companies. Only relatively recently has small-molecule screening emerged as a tool in academic biological research *(4, 5)*. Therefore, we have focused here on screening for molecules that are growth inhibitory to model bacteria, where the goal is to yield a rich supply of bioactive molecules that could be further characterized with mechanistic studies.

Below we describe methodologies outlining (i) the development of a sensitive and robust growth-inhibition assay; (ii) a screening campaign against a library of small molecules; and (iii) follow-up studies of active compounds with the target bacteria. Note that active compounds from the primary screen can be subjected to a number of different tests before moving to the more challenging task of identifying their cellular targets and their mechanism of action. Of these tests, we describe the determination of the dose-response relationship and the associated parameters MIC and EC_{50}.

1.2. Preamble to the Methodology Provided

The instructions herein describe how to screen a library of small molecules against a model microbe. For this purpose, the following assumptions have been made:

1. Any necessary subcloning of the microbe of interest has been completed.
2. Appropriate control assays have been established.
   (a). Negative control: a measure of the assay signal in the absence of any interfering small molecules;
   (b). Positive control: a measure of the assay signal in the presence of a small molecule with the desired effect (i.e., this effect could be a decrease or increase in assay signal, depending on whether inhibitors or activators, respectively, are sought) (*see* **Note 1**).

3. The assay has been adapted to microtiter plates.
4. An automated liquid handler will be used for all liquid transfers. Specific methods for such handlers are not included here.
5. Library compounds are dissolved in dimethylsulfoxide (DMSO).
6. The final concentration of compounds to be tested in each assay has been established. For microbial systems, typical concentrations range from 1 to 5 μM for natural products and known bioactive compounds, to 10–20 μM for diverse synthetic molecules.

For these instructions, the following story is used as an illustrative example. A strain of *E. coli* subcloned with a vector that expresses a gene conveying resistance to kanamycin is screened to identify novel growth inhibitors. The assay is performed in duplicate in separate 96-well clear microtiter plates with a final volume of 100 μL (2 μL of 1 mM tested compound or neat DMSO, plus 98 μL of bacterial culture), with 20 μM final compound concentration tested. The assay is monitored for bacterial growth, as determined by absorbance at 600 nm (i.e., A_{600}). Negative controls (100% growth) contain *E. coli* and neat DMSO; positive controls (0% growth) contain *E. coli* and a lethal drug, in this case ampicillin (100 μg/mL) or chloramphenicol (25 μg/mL) in DMSO. All assays contain 2% DMSO (v/v). Library compounds are housed in storage plates in neat DMSO within columns 2–11 only; columns 1 and 12 are empty. All liquid transfers are carried out with a Biomek FX (Beckman/Coulter) using disposable tips.

2. Materials

2.1. Preparation of Bacterial Culture

1. Appropriate frozen bacterial stock.
2. Sterile test tubes.
3. Luria-Bertani (LB) media: 10 g of bacto-tryptone, 5 g of bacto-yeast extract, and 10 g of NaCl per liter of deionized water. The media is distributed into 100–500 mL aliquots in screw-cap glass bottles or capped glass flasks, then sterilized for 45 min by autoclaving on liquid cycle. Once the media cool down, supplement it with, if applicable, appropriate concentration of drug used for selection (for this example, 50 μg/mL kanamycin) *(6)*.
4. Agar plates: 1.5% bacto-agar in LB media is sterilized by autoclaving for 45 min on liquid cycle. Before pouring the agar plates into Petri dishes, allow the media to cool to ~50°C, supplement it with any necessary drugs used for selection

(for this example, 50 μg/mL kanamycin), mix well by swirling, and then pour into Petri dishes. Allow a period of 20–30 min for the agar to harden. Any unused plates can be stored at 4°C *(6)*.

5. Sterile wooden stick (toothpick) or loop for colony picking.
6. Sterile test tubes or flasks for liquid culture.
7. Bunsen burner.
8. Temperature-controlled incubator.
9. Temperature-controlled shaker incubator.

2.2. Determination of Appropriate Incubation Time

1. Working bacterial culture as described in **Subheading 3.1., step 8a**.
2. Neat DMSO.
3. Appropriate antimicrobial compounds as negative controls: for this example, ampicillin (5 mg/mL) or chloramphenicol (1.25 mg/mL) in neat DMSO stored in aliquots at –20°C. If an appropriate antimicrobial is not available, 15 mL LB media is required.
4. A sterile, lidded, 96-well, flat-bottom, clear plate (Corning; Lowell, MA).
5. An appropriate detector: for this example, a plate reader that measures absorbance at 600 nm.
6. Temperature-controlled incubator.

2.3. Primary Screen

Note that all microtiter plates and tips must be compatible with the chosen liquid handler.

1. Working culture from **Subheading 3.1., step 8b**.
2. Kanamycin (50 mg/mL) stored in aliquots at –20°C.
3. Appropriate antimicrobial compounds as positive controls; for this example, ampicillin (5 mg/mL) or chloramphenicol (1.25 mg/mL) in neat DMSO stored in aliquots at -20°C. If an appropriate antimicrobial is not available, one 96-well, –2 mL high-profile plate and fresh LB media are required.
4. Neat DMSO. Use fresh daily, as this solvent is very hygroscopic. Store at room temperature.
5. Library compounds in 96-well storage plates (1 mM in DMSO). Store at –20°C. Take care to minimize exposure of compounds to air, as DMSO is very hygroscopic.
6. Sterile, lidded, 96-well microtiter plates appropriate for the assay: in this example, clear plates are used (Costar 96-well flat-bottom sterile polystyrene plates) (*see* **Note 2**).

7. A high-profile DMSO-resistant storage block (Corning 0.5 mL polypropylene plate).
8. Reservoirs that hold ~300 mL of reagent (Nalge Nunc International; Rochester, NY);
9. Twenty-microliter, 96-racked tips for the liquid handler (Molecular BioProducts: San Diego, CA) (*see* **Note 3**);
10. Two hundred-microliter, 96-racked tips for the liquid handler (Beckman Coulter; Fullerton, CA).
11. Humidity- and temperature-controlled incubator (*see* **Note 4**);
12. Spray bottle containing 95% ethanol;
13. Water with a minimum resistivity of 17.5 MΩ cm (hereafter referred to as "water").

2.4. Follow-Up on Primary Actives: Dose-Response Determination

1. Any necessary drugs used for selection (for this example, 50 μg/mL kanamycin);
2. LB media from **Subheading 2.1., step 2**.
3. Working culture from **Subheading 3.1., step 8c**.
4. Neat DMSO.
5. Appropriate antimicrobial for positive controls: for this example, ampicillin (5 mg/mL) or chloramphenicol (1.25 mg/mL) in neat DMSO stored in aliquots at –20°C. If an appropriate antimicrobial is not available, one 96-well, 2-mL, high-profile plate and fresh LB media are required.
6. Compounds identified as actives from **Subheading 3.3**, at 5 mM, or the highest concentration available.
7. Low-volume polypropylene plates to reformat active compounds.
8. Sterile, lidded, 96-well microtiter plates appropriate for the assay: in this example, clear plates are used (Costar 96-well flat-bottom sterile polystyrene) (*see* **Note 2**).
9. Reservoirs that hold ~300 mL of reagent (Nalge Nunc International).
10. Twenty-microliter 96-racked tips for the liquid handler (Molecular BioProducts) (*see* **Note 3**).
11. Two hundred-microliter, 96-racked tips for the liquid handler (Beckman Coulter).
12. Humidity- and temperature-controlled incubator (*see* **Note 4**).
13. Spray bottle containing 95% ethanol.
14. Water.

3. Methods

3.1. Preparation of Bacterial Culture

1. Next to a flaming Bunsen burner, open the frozen strain stock, scrape up a small amount using a sterile loop or toothpick, and streak it on an agar plate (*see* **Note 5**).
2. Incubate plate overnight at 37°C. This plate can be used over 7 days to pick colonies.
3. The next day, next to a flaming Bunsen burner, inoculate 5 mL of LB media in a sterile test tube with a single colony from the agar plate using a sterile toothpick (*see* **Note 6**).
4. Grow the liquid culture overnight in the incubated shaker at 37°C, 250 rpm (*see* **Note 7**).
5. The next day, next to a flaming Bunsen burner, dilute the overnight culture 1:100 into fresh LB media. Typically, this dilution is 30 μL into 3 mL in a sterile test tube. This is the *subculture* (*see* **Note 8**).
6. Place the subculture in the incubated shaker at 37°C, 250 rpm, and grow until it reaches mid-log phase (A_{600} = 0.4 – 0.5): for this example, ~3 h (*see* **Note 9**).
7. Next to a flaming Bunsen burner, prepare the *working bacterial culture* by diluting the subculture 1:10,000 into one of the screw-cap bottles or capped flasks containing LB media and mix well.
8. The number and volume of working bacterial cultures to be prepared is dependent on the procedure to be completed next:
   a. If determining growth/signal over time, one flask of 100 mL culture is sufficient.
   b. If undertaking a primary screen, three separate 500 mL working bacterial cultures are needed each day. Inoculate each culture, as in **step 7**, from three separate subcultures, as in **step 5**, that have been prepared every 2 h; this step will ensure that fresh working culture is used throughout the day (*see* **Note 10**).
   c. If undertaking follow-up studies, the number of cultures and volumes needed will vary with the number of compounds to be tested. Follow method as in **Subheading 3.1., step 8b**, preparing one 500 mL bacterial culture for every 48 assay plates.

3.2. Determination of Appropriate Incubation Time

If an appropriate antimicrobial for positive controls is available (*see* **Subheading 2.2.**, item 3), follow **steps 1–5**. Otherwise, start with **step 6**.

1. Prepare a working bacterial culture as in **Subheading 3.1., step 8a**.
2. Using either a multichannel pipettor or a liquid handler, aliquot 98 μL of the culture to two 96-well plates.
3. Add neat DMSO to one half of the wells of each plate (negative controls yield 100% growth). The final concentration of DMSO is the same as what will be used in the primary screen (typically, 0.2–5%).
4. Add the antimicrobial reagent to the other half of each plate (positive controls yield 0% growth). Ensure the same final concentration of DMSO is used as in the previous step.
5. Proceed to **step 9**.
6. Prepare a working bacterial culture as in **Subheading 3.1., step 8a**.
7. Using either a multichannel pipettor or a liquid handler, aliquot 98 μL of the culture to one half of the wells of two 96-well plates (positive controls) and 98 μL of LB media to the other half (negative controls).
8. Add neat DMSO to all wells of each plate. The final concentration of DMSO is the same as what will be used in the primary screen (typically, 0.2–5%).
9. Measure the absorbance at 600 nm, and any other signal as appropriate, for each plate. Note that if it is necessary to measure a signal in end point fashion, quench three positive controls and three negative controls for each measurement.
10. Incubate both plates at the appropriate temperature (e.g., 37°C). Take absorbance and any other necessary readings every 2 h or as manageable, until the growth of the microbe reaches the stationary phase, as indicated by little comparable change in growth over several hours.
11. Plot growth/signal versus time. Set the most appropriate *incubation time* for use in the primary screen by choosing the time at which microbial growth is still in log phase, and any secondary signals are maximized (*see* **Note 11**). At the selected incubation time, the calculated statistical parameter Z' should be >0.5 if looking to identify target inhibitors in the primary screen (*see* **Notes 12** and **13**).

3.3. Primary Screening

3.3.1. Labware Preparation

1. Allow library compounds to come to room temperature while still sealed in the storage plates. Unseal compounds immediately prior to use and reseal as soon as possible after use to minimize uptake of atmospheric water (*see* **Note 14**).
2. Label all assay plates appropriately; the number of the compound plate to be tested and replicate number are usually sufficient (*see* **Note 15**). If using an antimicrobial agent

for negative controls, follow **steps 3–5**. Otherwise, skip to **step 6**.

3. Add neat DMSO to the positive and negative control wells of each assay plate in the same volume as the library compounds for testing (in this example, 2 μL). To the 0.5 mL DMSO-resistant storage block, add neat DMSO to wells A1, C1, E1, G1, B12, D12, F12, and H12 in an appropriate volume for at least several runs of the liquid handler (*see* **Note 16**). Add the same amount of ampicillin (5 mg/mL) or chloramphenicol (1.25 mg/mL) in DMSO to wells B1, D1, F1, H1, A12, C12, E12, G12 (*see* **Note 17**). This is the *control block*.
4. Add an appropriate volume of microbe to one of the 300 mL reservoirs.
5. Skip to **step 8**.
6. Add neat DMSO to the positive and negative control wells of each assay plate in the same volume as the library compounds for testing (in this example 2 μL). To the 0.5 mL DMSO-resistant storage block, add neat DMSO to columns 1 and 12 in an appropriate volume for at least several runs of the liquid handler. This is the *control block* (*see* **Notes 16** and **17**).
7. Add LB media to wells B1, D1, F1, H1, A12, C12, E12, and G12 of the 2 mL high-profile plate. Add working bacterial culture to all other wells.
8. Add 300 mL water to the second 300 mL reservoir.

3.3.2. Liquid Handling

Note that a full method on the liquid handler from start to finish without any user intervention is referred to as a "run." Using the automated liquid handler, transfer the appropriate volumes of microbe and compounds to the assay plates, using the following outline as a guide.

1. The number of assay plates that are prepared for each run of the liquid handler will be dictated by the size of the working surface of the instrument. For any liquid handler, each run should be able to access (*see* **Notes 18** and **19**):
   - *n* compound storage plates
   - 2*n* assay plates
   - The control block
   - *n* 20-μL racked tips
   - 1 × 200-μL racked tips
   - 1 × 300-mL reservoir, or 2-mL high-profile plate containing microbe culture
   - 1 × 300 mL reservoir containing water

2. Add reagents to the assay plates in the following order:
   a. Use the 20 μL liquid-handling tips to transfer the appropriate amount of library small molecules (in this example, 2 μL) from compound storage plate 1 to assay plate 1, replicate 1 (or R1) and then assay plate 1, R2. Using this same box of tips, transfer the same volume from the control block to these same two assay plates in the same order (*see* **Note 20**). Discard tips (*see* **Note 21**).
   b. Repeat **step 2a** to distribute all compound plates to the appropriate assay plates for this run.
   c. Using one box of 200 μL tips, aliquot the screening broth (98 μL) to each assay plate in the following sequence:
      - Assay plate 1, R1 (*see* **Notes 22** and **23**)
      - Wash tips in 300 mL water in reservoir (*see* **Note 24**)
      - Assay plate 1, R2
      - Wash tips in 300 mL water in reservoir
      - Assay plate 2, R1
      - Wash tips in 300 mL water in reservoir
      - Assay plate 2, R2
      - Wash tips in 300 mL water in reservoir

      Use the same 200 μL tips for approximately ten assay plates, or at least one run, then discard.
3. Replace the lids of the assay plates in the reverse order to that in which they were removed. Remove assay plates from the liquid handler and stack.
4. Prepare for the second liquid-handling run by replenishing reagents as needed, and using new compound and assay plates and appropriate tips. Start the second run.
5. While the liquid-transfer of **step 4** is occurring, if applicable, read the assay plates on the appropriate detector for a $t = 0$ measurement (*see* **Note 25**). Note the time at which measurements are made (*see* **Note 26**).
6. While the liquid-transfer of **step 4** is occurring, transfer the prepared assay plates to an incubator at the appropriate temperature and humidity for the microbe (*see* **Note 27**). Note the time that each stack was placed in the incubator (*see* **Note 28**).
7. Repeat **steps 2–6**, until approximately 2 h has elapsed; a reasonable throughput is 12 compound plates (960 compounds in duplicate) per hour.
8. Continue the primary screen as above, using a fresh working culture every 2 h.

9. At the end of each day, reseal all compounds and store at –20°C.
10. Incubate the microbes for the appropriate incubation time as determined in **Subheading 3.2.**, then measure the appropriate signal, e.g., growth as determined by A_{600}.
11. Calculate the percent growth/signal of each assay by

$$\%G = \left(\frac{S - \mu_{+c}}{|\mu_{-c} - \mu_{+c}|} \right),$$

where S is the signal for either a positive, negative, or test reaction, and μ_{+c} and μ_{-c} are the average S for the positive and negative controls, respectively. Note $\%G$ for each tested compound is to be calculated using the control wells contained within the same assay plate.

12. Identify primary active compounds; typically, these are identified as those compounds that result in growth/activity 3 standard deviations below or above (for inhibitors or activators, respectively) the average of the negative controls (i.e., 100% activity).

3.4. Follow-Up Tests of Primary Actives: Dose–Response Determination

1. Prepare the working bacterial culture as in **Subheading 3.1., step 8c**.
2. Reformat active compounds from **Subheading 3.3** into polypropylene plates: with one compound per well, aliquot 15 μL of 8 compounds into wells A1–H1, and 10 μL neat DMSO to all wells in columns 2–11. Using a multichannel pipettor or liquid handler, transfer 4.7 μL of the compounds in column 1 into the DMSO into column 2. Continue the serial dilution across the plate to column 11 (*see* **Note 29**).
3. Aliquot 15 μL neat DMSO to wells A12–D12.
4. Aliquot 15 μL of the appropriate antimicrobial (e.g., ampicillin (5 mg/mL) or chloramphenicol (1.25 mg/mL)) into wells E12–H12. If an antimicrobial is not available, aliquot neat DMSO to these wells.
5. If an appropriate antimicrobial agent (*see* **Subheading 2.4., step 5**) is available, add the working bacterial culture to the 300 mL reservoir. If not, add LB media to wells E12–G12 of the 2-mL high-profile plate. Add working bacterial culture to all other wells.
6. Aliquot compounds and bacteria to assay plates as in **Subheading 3.2., steps 2–11**, ignoring the comment regarding the control block in **step 2**.
7. Plot percent growth/signal against compound concentration and fit to:

$$\%\text{Growth} = \frac{R}{1+\left(\frac{[I]}{EC_{50}}\right)^{s}} + B,$$

where R is the extrapolated percentage growth in the absence of inhibitor, $[I]$ is the concentration of tested compound (μM), s is the slope (or Hill factor), and B is background signal, i.e., the extrapolated percentage growth at which the inhibitor exerts its maximal effect. Define the MIC for each compound as the lowest concentration at which no visible growth of the strain is observed (*see* **Note 30**).

4. Notes

1. If inhibitors are sought in the screen, an adequate positive control would be to use bacteria-free media (i.e., exclude the target from the assay), or to include a known antimicrobial agent. If activators are sought, it is not strictly necessary to have a positive control.
2. It is preferable to use clear, black, or white plates for absorbance, fluorescence, and luminescence detection, respectively. It is possible to use black or white plates with clear bottoms to enable the detection of both absorbance and either fluorescence or luminescence.
3. If you include antimicrobials in the screening broth, it is not strictly necessary to use sterile tips.
4. It is preferable to use an incubator that maintains humidity above 80%; this helps in preventing evaporation from assay plates during incubation.
5. Cryoprotectants such as glycerol or DMSO are used for freezing and long-term storage of cells to prevent any damaging effects caused by the formation of ice crystals. Frozen glycerol stocks are made by mixing an aliquot of the overnight culture with an equal volume of sterile 30% glycerol in LB media, and storing in cryovials at –80°C. Alternatively, frozen DMSO stocks are made by mixing 1 mL of overnight culture with 90 μL of freezing solution (DMSO and 10% ethanol) and storing in cryovials at –80°C.
6. Using a single colony guarantees cell homogeneity. For a more stringent condition of homogeneity, a single colony is picked from the plate and streaked on a second plate from which the liquid culture would finally be inoculated with a single colony.

7. The conditions describe requirements for wild-type *E. coli* to reach stationary phase when grown in rich LB media; the length of incubation needed for a culture to reach stationary phase is dependent on the strain of interest and growth media used. A growth profile of the strain of interest should performed in order to establish these conditions (*see* **Subheading 3.2**).
8. Subculturing serves to bring the cells to a phase of exponential growth.
9. If the 3 mL subculture is found to reach mid-log phase too slowly, the subculture can be set up in a flask with a better surface area-to-volume ratio (e.g., 500 μL overnight culture into 50 mL LB media in an Erlenmeyer flask) to increase the growth rate.
10. If the screen is being run all day, then several subcultures should be set up from the overnight culture at regular time intervals. This ensures that the later assay plates are not set up with an overgrown screening broth. For this purpose, the subcultures need to be staggered a few hours apart (**Table 1**).
11. Generally, when adding microbes to assay plates with test compounds, it is desirable to use a culture concentration in which the organism is in lag phase or early log phase. At the time of final detection, after an incubation period, the microbes should be in log phase; if the stationary phase is reached, the assay will be much less sensitive to subtle effects of small molecules tested.
12. Z' is a simple, dimensionless statistical measure of the quality of a small-molecule screening campaign. It is defined as:

$$Z' = 1 - \left(\frac{3\sigma_{+c} + 3\sigma_{-c}}{|\mu_{+c} - \mu_{-c}|} \right),$$

where σ_{c+}, σ_{c-}, μ_{c+}, and μ_{c-} are the standard deviations (σ) and means (μ) of the positive (c+) and negative (c−) controls *(7)*. A Z' value of 0.5 or greater represents a set of data with

Table 1
Establishment of working bacterial cultures for primary screening throughout the day

Procedure	1	2	3
Subculture set up at	8 a.m.	10 a.m.	12 noon
Working bacterial culture set up at	11 a.m.	1 p.m.	3 p.m.

a good screening window and clear separation between the positive and the negative controls.

13. *See* **Fig. 1** for an example of a growth curve for *E. coli*. Twenty hours was chosen as the incubation time for the primary screen, as the bacteria were still in log phase at this time and the calculated Z' was greater than 0.5.
14. It is good practice to spin compound plates in a centrifuge to prevent compound loss when unsealing the plates.
15. Make the label short, mark whichever side is still legible when on the liquid handler, and, preferably, keep the legible side facing outwards when in the incubator for easy identification; it is not necessary to label lids.
16. It is suggested that a 0.5 mL block be used, as this volume should be sufficient to hold the total volume of DMSO needed for a day's screening.
17. By alternating positive and negative controls in this manner, it is easier to spot well-specific variations in each assay plate.
18. *See* **Fig. 2** for an example of how to organize the working surface of the Biomek FX for this example.
19. Spray the bench surface in front of the liquid handler with 95% ethanol and wipe down. Remove the lids from the sterile 96-well plates by beginning with the row farthest from the operator, and working towards the operator; this minimizes hand movement over open plates, a potential source of contamination.
20. Recall that the control block contains reagent in columns 1 and 12 only, whereas the same columns in the compound storage plates are empty.

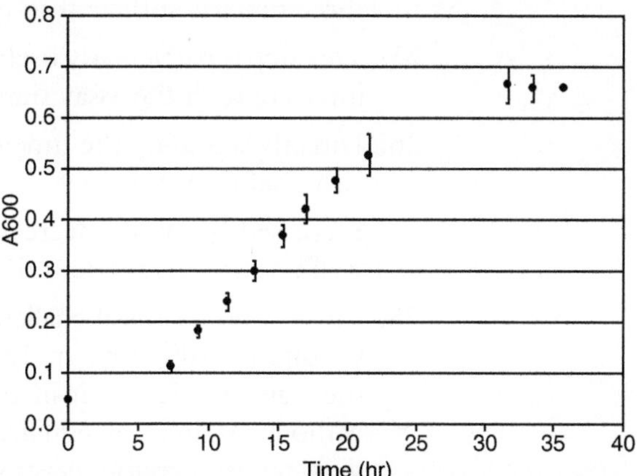

Fig. 1. *E. coli* growth over time in the presence of 5% DMSO. Data shown are the average of 96 replicates.

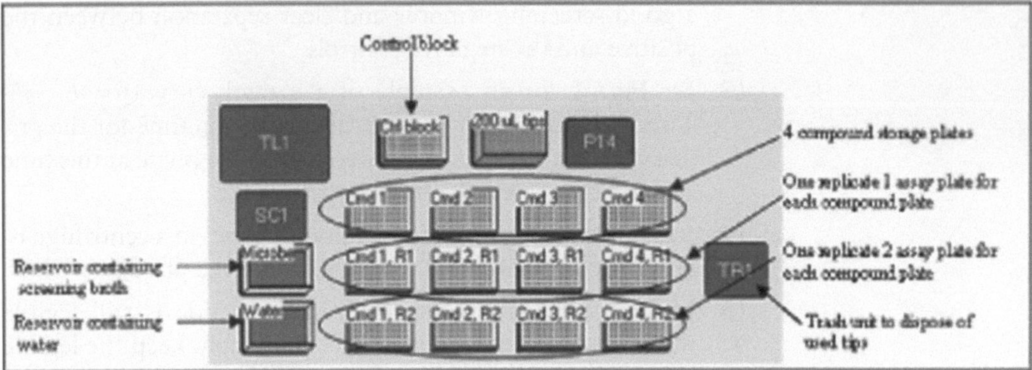

Fig. 2. One way in which to organize the working surface of the Biomek FX for a primary screen. Not shown is storage space for 20-μL tips for compound and control block transfer; these tips are introduced to the liquid handler by way of a shuttle (SC1) from the storage carousels. TL1 is the position where disposable tips are loaded onto the 96-channel pipetting head. One extra space on the liquid handler, P14, remains empty throughout each run.

21. Add compound to assay plates first so the same tips can be used for addition to both replicates; if the bacteria were in the plate first, a second box of tips would be required to avoid contaminating the compounds. If available, pin tools can be used to add compounds directly to microbes in assay plates.
22. It is recommended that the microbe is mixed in its storage reservoir at the start of each run by aspirating and dispensing 150 μL four times to prevent solution inhomogeneities from cell settling.
23. It is recommended that the screening broth be aspirated and dispensed several times after addition to each assay plate to ensure adequate mixing with compound.
24. Washing (i.e., aspirating and dispensing) with 150 μL water three times is sufficient to prevent compound carryover.
25. This step is particularly useful to flag compounds that directly interfere with the assay signal.
26. Directly stamping the time in detector output files is particularly useful for this step.
27. Increased humidity decreases media evaporation leading to well position-dependent data (i.e., edge effects).
28. There is no technical challenge to incubating assay plates with shaking (usually done at 150 rpm). However, we found that the standard deviation in the growth of microbes incubated without shaking was significantly lower than those grown with shaking; we therefore never shake during incubation time.
29. This dilution factor will result in 11 compound concentrations that span five logs at half-log intervals.

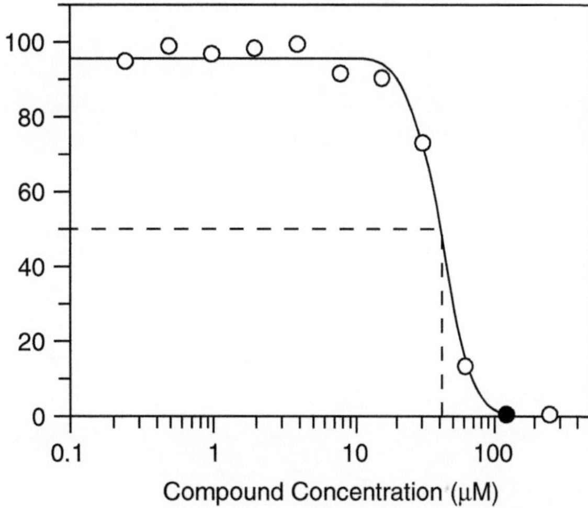

Fig. 3. Illustrative example of a dose–response relationship between *E. coli* and an active compound identified in a primary screen. The *solid line* shows the best fit of the data to a four-parameter equation to determine EC_{50} (*see* **Subheading 3.4., step 7**). The calculated EC_{50} value (41.0 μM) is highlighted by the *dashed line*. The single solid data point was defined as the MIC (125 μM).

30. Although both the MIC and EC_{50} values of a compound are important measures of its bioactivity, the latter value is considered a more precise parameter by which to measure compound potency (**Fig. 3**). Values of EC_{50} are calculated from the entire dose–response data set and are not influenced by the selection of compound concentrations to determine the point of complete growth inhibition of the microbe, as with MIC values.

References

1. Hayashi, K., Morooka, N., Yamamoto, Y., Fujita, K., Isono, K., Choi, S., et al. (2006) Highly accurate genome sequences of Escherichia coli K-12 strains MG1655 and W3110. *Mol. Syst. Biol.* 2, 2006.0007.
2. Baba, T., Ara, T., Hasegawa, M., Takai, Y., Okumura, Y., Baba, M., et al. (2006) Construction of Escherichia coli K-12 in-frame, single-gene knockout mutants: the Keio collection. *Mol. Syst. Biol.* 2, 2006 0008.
3. Saka, K., Tadenuma, M., Nakade, S., Tanaka, N., Sugawara, H., Nishikawa, K., et al. (2005) A complete set of Escherichia coli open reading frames in mobile plasmids facilitating genetic studies. *DNA Res.* 12, 63–68.
4. Brown, E. D. (2003) Screening in academe: a perspective on implementation of university-based small molecule screening. *J. Biomol. Screen.* 8, 377–379.
5. Stein, R. L. (2003) High-throughput screening in academia: the Harvard experience. *J. Biomol. Screen.* 8, 615–619.
6. Sambrook, J. and Russell, D. (ed.) (1989) Molecular Cloning: A Laboratory Manual. Cold Spring Harbor Laboratory Press, Woodbury, NY.
7. Zhang, J. H., Chung, T. D., and Oldenburg, K. R. (1999) A simple statistical parameter for use in evaluation and validation of high throughput screening assays. *J. Biomol. Screen.* 4, 67–73.

Chapter 3

Cell-Based Assays to Probe the ERK MAP Kinase Pathway in Endothelial Cells

Michael R. Wyler, Deborah H. Smith, Eftihia Cayanis, Udo Többen, Nathalie Aulner, and Thomas Mayer

Summary

To understand signaling pathways in mammalian cells, cell-based assays are relatively new and extremely powerful tools. We have developed a battery of phenotypic assays to study signaling; two of them are described in detail in this chapter. A subset of these assays monitors mitogen-activated protein (MAP) kinase pathways. MAP kinases are principal regulators of fundamental processes in mammalian cells, including growth, cell division, differentiation, stress responses, and neoplastic transformation. Here we describe two cell-based assays querying the function of ERK (extracellular signal regulated kinase), one of the three principal MAP kinases in mammalian cells. We selected human umbilical vein endothelial cells (HUVECs), a primary cell type, because they show a very dynamic response to various activators. Both assays are phenotypic assays and use well-established phosphorylation-specific primary antibodies to study activation. Fluorochrome-coupled secondary antibodies were used to label phosphorylated target proteins; images were captured with the INCell Analyzer 3000 and analyzed with the INCell Analyzer 3000 software. The first of these two assays monitors phosphorylation of ERK1/2, while the second assay monitors activation of the transcription factor CREB (cAMP response element-binding protein). The assays described in this chapter cover major checkpoints of the ERK signaling pathway: (1) MAP kinase activation and (2) subsequent transcription factor activation. Both assays exhibit robust performance and can easily be used for high-throughput screening.

Key words: Automated microscopy, Cell-based assays, HUVEC, Phosphorylation-specific antibodies, Primary cells, Signal transduction.

P.A. Clemons et al. (eds.), *Cell-Based Assays for High-Throughput Screening, Methods in Molecular Biology, Vol. 486*

DOI: 10.1007/978-1-60327-545-3_1

1. Introduction

Defects in the complex cellular signaling network that regulates cell proliferation and survival can result in neoplastic transformation of the cell. These defects have been used as models for the complex collection of diseases known as cancer. A major part of the signaling network relevant to proliferation and survival comprises three mitogen-activated protein (MAP) kinase pathways, termed the extracellular signal regulated kinase (ERK), JNK, and p38 pathways, so named for the principal kinases *(1–3)*. The significance of these regulatory pathways is underlined by recent and ongoing development of a number of MAP kinase inhibitors as cancer therapeutics *(4–6)*. High-throughput cell-based assays have proven to be powerful tools to deconvolute complex signaling pathways, including the MAP kinase pathways. For many decades, in vitro culture of mammalian cells has been harnessed to study basic biological questions. Diverse transformed and primary mammalian cells have been exploited to study diseases, including diabetes, cancer, and neurodegenerative disease *(7, 8)*. However, many breakthroughs have depended on the serendipitous discovery of a small-molecule inhibitor with specificity for a particular biological process. Recent technological advances in high-throughput screening have reduced the dependence of progress on serendipity. Existing and newly developed hardware, such as automated microscopes, imaging platforms, and robotic accessories, as well as software, such as automated image analysis programs, have enabled the probing of larger segments of chemical space for new biological activities. Here, we describe the application of cell-based assays to investigate the ERK pathway.

Many features of cell growth, development, and survival are controlled by MAP kinase pathways. Although the focus here is on the ERK pathway, many of the techniques can easily be adapted to the other two MAP kinase pathways. The ERK pathway is generally activated by a mitogen binding to a receptor tyrosine kinase, but this pathway can also be activated by ion channels, G-protein-coupled receptors, and integrins *(9, 10)*. An activated receptor is linked to a cascade of MAP kinases by a specific adaptor molecule, such as protein kinase C (PKC), son of sevenless (SOS), or the small GTPases of the ras/rap family. For the ERK pathway the cascade begins with phosphorylation of raf (a MAP kinase kinase kinase), followed by MEK1/2 (MAP kinase kinase) and ERK1/2 (MAP kinase). Phosphorylated ERK translocates to the nucleus where its role is to effect transcriptional activation and repression.

Successful development of a high-throughput cell-based assay poses several challenges. In general, owing to the large scale of the enterprise, considerations of biological authenticity must be balanced against cost and practicality. For example, regarding choice of cell type, it is more practical to use an established cell line, although primary cells may be more biologically relevant. For the assays described here, we have chosen human umbilical vein endothelial cells (HUVEC), which are neither transformed nor genetically manipulated but can be cultivated for at least 16 cell doublings in culture *(11)*. Furthermore, they are commercially available in lots from pooled individuals, reducing genetic variability.

Another critical parameter of every assay is the type of readout that is used. For a cell-based assay focused on a signaling pathway, there is often considerable flexibility in the choice of readout. The most accessible readouts for interrogation of a signaling pathway can be any step in which a protein is modified (e.g., phosphorylated), translocated (e.g., moves from cytoplasm to nucleus), or changed in its abundance, provided the appropriate antibody reagents are available *(12)*. An alternate approach is to generate a reporter molecule that can monitor a given step of the pathway. This could be a GFP-tagged protein to monitor translocation, or a promoter/enhancer driving the expression of luciferase to monitor transcriptional activation *(13, 14)*. The use of reporter constructs involves extensive manipulation of cells that may affect the pathway to be studied.

We describe here two assays that monitor different stages of the ERK1/2 signaling pathway. In both assays, the pathway is activated using a phorbol ester, TPA (phorbol-12-myristate-13-acetate). TPA treatment is a standard procedure used to directly activate PKC, thereby bypassing the cell-surface receptor. This treatment produces a more robust and reproducible response than activation via a cell-surface receptor. The first assay described here detects phosphorylation and nuclear translocation of ERK1/2 using antibodies specific for the activated form of phospho-ERK. The second assay monitors a downstream event, namely the phosphorylation of the transcription factor CREB at Ser 133.

2. Materials

2.1. Cell Culture

1. Activators/inhibitors: Phorbol 12-myristate 13-acetate (TPA/PMA) (Calbiochem; San Diego, CA) and U0126 (Calbiochem).
2. Cells: HUVECs obtained as frozen stocks (Lonza Biosciences; Basel, Switzerland).
3. Medium: EGM-2 medium with supplements (Lonza Biosciences).

4. Phosphate buffered saline (PBS) for tissue culture (Lonza Biosciences).
5. DMEM containing 10% fetal calf serum (Lonza Biosciences).
6. T175 flasks (Corning; Lowell, MA).
7. View plates (PerkinElmer; Waltham, MA).
8. 0.05% Trypsin/EDTA (HyQ, Hyclone; Logan, UT).
9. Anti-phospho-ERK1/2 (Thr202/Tyr204) (BDBiosciences; San Jose, CA).
10. Anti-phospho-CREB (Ser133) (Cell Signaling; Danvers, MA).
11. Goat anti-mouse coupled to Alexa 647 nm (Molecular Probes; Carlsbad, CA).

2.2. Fixing and Staining

1. Permeabilizing/blocking solution: 10% (v/v) newborn calf serum (Gibco/Invitrogen; Carlsbad, CA; *see* **Notes 1** and **2**), 0.2% (w/v) Triton X-100 in PBS.
2. Wash solution: 10% (v/v) newborn calf serum (Gibco/Invitrogen) in PBS.
3. Primary antibody solutions: 0.25 µg/mL mouse IgG against ERK1/2 in PBS containing 10% (w/v) newborn calf serum and 0.2% Triton X-100.
   - The anti-CREB (Ser 133) antibody stock solution (purified mouse IgG) was diluted 1:1,000 in PBS containing 10% (w/v) newborn calf serum and 0.2% Triton X-100.
4. Secondary antibody solution: 2 µg/mL anti-mouse antibody coupled to Alexa Fluor® 647 nm (Molecular Probes) and 1µg/mL Hoechst 33342 (Calbiochem) in PBS containing 10% (v/v) newborn calf serum and 0.2% Triton X-100.

2.3. Image Acquisition

2.3.1. Equipment

1. Automated Microscope: INCell Analyzer 3000 (GE Healthcare; USA).
2. Plate Hotel: Kendro Cytomat Hotel (Kendro; Germany).
3. Robot Arm: Mitsubishi MELFA RU-2AJ arm (Mitsubishi; Japan).

2.3.2. Camera and Hardware Settings

1. Blue Camera: Neutral Density Filter 1, Emission filter 450–25 nm.
2. Red Camera: Neutral Density Filter 0, Emission filter 695–55 nm.
3. Exposure time of 1.7 ms.
4. Binning of 1.

3. Methods

Methods that are common to the assays described in this article are listed in the beginning of the Methods section. To simplify assay development and implementation we have developed standardized cell-culture and -staining procedures for our cell-based assays. These standardized procedures are described in this section. Although we routinely develop assays in a 96-well format, we have found that most assays can be easily transferred to a 384-well format if a higher-density format is needed.

3.1. Cell Culture

1. HUVECs were obtained as pooled, frozen stocks and grown in EGM-2 medium with supplements.
2. Cells were grown at 37°C, 5% CO_2, and 100% humidity (standard conditions) in T175 flasks in an Autoflow incubator (Nuaire).
3. To expand HUVECs, the medium was aspirated and the monolayer rinsed once with 20 mL PBS, followed by the addition of 5 mL of a trypsin/EDTA solution (0.05% trypsin/0.02% EDTA in PBS).
4. Cells were detached by incubation with 5 mL of trypsin/EDTA solution per T175 flask for 6–10 min, then resuspended in 20 mL of DMEM containing 10% fetal calf serum, and centrifuged for 5min at 1,000×g.
5. HUVECs were *see* ded at dilutions of 1:2–1:3 and grown to confluence. Experiments during the developmental phase were performed with cells kept in culture up to passage 9.

3.2. Cell Seeding and Stimulation

1. HUVECs were seeded using a 12-channel automatic pipettor at a density of 10,000 cells per well in 96-well plates ("View-Plates") and grown for 16 h under standard conditions.
2. Specific signaling was initiated by removing the growth medium and replacing it with 100 μL prewarmed medium containing the activating agents.

3.3. Fixing and Staining

To facilitate easy transition from the assay development phase to different automated liquid-handling systems, we drastically reduced washing steps and incubation times for the postfixation procedures.

3.3.1. Fixation

1. Add 35 μL 16% paraformaldehyde to each well (*see* **Note 3**).
2. Fix for 20 min at room temperature.

3. Remove fixative solution and wash once with 100 μL PBS.
4. Store cells in 100 μL PBS at 4°C until ready to stain.

3.3.2. Staining (All Steps Are Performed at Room Temperature)

1. Remove PBS from cells.
2. Add 100 μL permeabilizing/blocking solution.
3. Incubate for 20 min.
4. Remove permeabilizing/blocking solution.
5. Add 50 μL primary antibody solution (*see* **Note 4**).
6. Incubate for 20 min.
7. Remove antibody solution and wash once with 175 μL wash solution.
8. Add 50 μL secondary antibody solution (*see* **Note 4**).
9. Incubate for 20 min.
10. Remove antibody solution and wash three times with 100 μL PBS.
11. Seal and store at 4°C in fresh 100 μL PBS until ready to scan.

3.4. Image Acquisition

1. Software settings ("Nuclear Traffic Module" of INCell Analyzer 3000 software).
   (a) Primary Measurement Parameters:
   - Nuc Marker: Blue
   - Cytoplasm: Red
   - Cyt Thresh: T=10
   - Clip Rings: N
   - Signal: Red
   - Sig Thresh: None
   - Erosion: 2
   - Dilation: 4

   (b) Object Definition Parameters:
   - Threshold: background+C
   - Constant C: 100.00
   - Discard Percentile: 10.00
   - Fill Holes: Y
   - Erosion: 0
2. Plates were imaged by sequential scanning using the red and blue channel.
3. To acquire the first image, samples were excited with the 647 nm laser line and images were recorded with the 695–55nm emission filter (red channel).
4. For the second image, the Hoechst dye was excited at 364 nm and images recorded with the 450–25nm emission filter (blue channel).

5. As standard procedure, we recorded a single square field (0.561 mm^2) from each well. Using an exposure time of 1.7 ms and a binning factor of 1, we averaged 200–300 cells per image.
6. After scanning the plates, data files including images and descriptive metadata files ("run files") were transferred from the internal hard drive to a file server.
7. Analysis of the images was performed on personal computers (operating system Microsoft XP Professional, Microsoft Corporation) using the INCell Analyzer software (GE Healthcare). Analysis parameters and analysis modules used for the assays described in this chapter are listed below.
8. Files containing the metadata were imported into Microsoft Excel to facilitate graphic representation of the results.

3.5. Two Assays to Monitor the ERK1/2 Pathway

3.5.1. ERK1/2 Phosphorylation Assay

After activation with the phorbol ester TPA, the MAP kinase ERKl/2 is phosphorylated at positions threonine 202 and tyrosine 204, respectively. Phosphorylated ERK forms a dimer that is translocated to the nucleus (**Fig. 1A**). Using phosphorylation-specific primary antibodies and secondary antibodies coupled to the fluorochrome Alexa 647, only the phosphorylated/activated form of ERK is detected/labeled (**Fig. 1B**). The nucleus is labeled by Hoechst 33342. As the readout we determine the average intensity per pixel arising from the phosphorylated ERK in

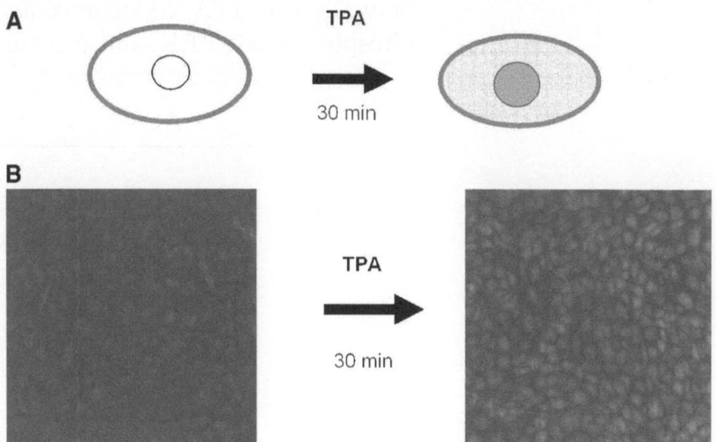

Fig. 1. Activation of ERK1/2. (A) Activation and translocation of phosphorylated ERK1/2 after TPA treatment. (B) HUVECs were stimulated with TPA as described in the text. To detect distribution of the MAP kinase ERK1/2, cells were fixed and permeabilized, and then incubated initially with an antibody against the phosphorylated form (Thr202/Tyr 204) of ERK1/2 and subsequently with secondary antibodies coupled to Alexa Fluor® 647. The left panel shows images of nonstimulated cells and the right panel shows images of TPA-stimulated cells.

the nucleus of the HUVEC. To analyze the images, the "Nuclear Traffic Module" of the INCell Analyzer 3000 software was used.

1. Cells were seeded at 10,000 cells per well and incubated for 16 h in EGM-2 medium using standard conditions. We found that 16 h incubation resulted in a constant and reproducibly low baseline of background ERK phosphorylation.

2. To activate the pathway, the medium was removed and 100 μL of EGM2 medium containing 1μM TPA was added.

3. After 30 min incubation at standard conditions, cells were fixed by adding 35 μL of 16% paraformaldehyde (*see* **Note 3**). We have observed that 30 min incubation with TPA works very well for most assay analysis requiring phosphorylation and/or nuclear translocation events. This includes assays for p38 and JNK MAP kinase activation, NFκB nuclear translocation, and the CREB phosphorylation/nuclear translocation assay described later in this chapter. **Fig. 1B** shows images of HUVECs treated with (**Fig. 1B**, right panel) or without TPA (**Fig. 1B**, left panel).

4. Pictures were taken with the INCell Analyzer 3000. Cells were immunostained with the phosphorylation-specific antibodies as described above. **Fig. 2** shows quantitative data obtained from analyses of images from the INCell 3000 software..

5. Increasing concentrations of TPA result in an increased phosphorylation and nuclear accumulation, and thus activation of ERK1/2 (**Fig. 2**). The curve shows a strong correlation between the TPA concentration and nuclear intensity of phosphorylated ERK, and plateaus at 60 nM with an AC_{50}

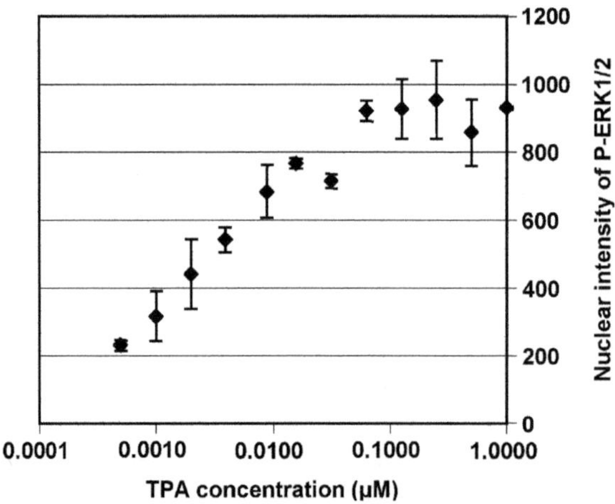

Fig. 2. TPA inducing phosphorylaton of ERK1/2 in a dose-dependent manner. HUVECs were stimulated for 30 min with the indicated concentrations of TPA. Cells were treated as described for Fig. 1. ERK1/2 activation is represented by nuclear fluorescence intensity (in arbitrary units) as a function of TPA concentration.

of 4nM. We have also successfully used H_2O_2 and sorbitol to stimulate phosphorylation and nuclear translocation of ERK1/2 (data not shown). These assays measure the response of the HUVECs toward oxidative and osmotic shock, respectively.

6. Activation of ERK strongly depends on MAP kinase kinase MEK1 activity as demonstrated with specific inhibitor U0126 *(15)*. U0126 inhibits the accumulation of phosphorylated ERK1/2 with an IC_{50} of 80 nM (**Fig. 3**). The results shown above validate our assay: ERK1/2 activation is dependent on TPA and is mediated by MEK1.

3.5.2. CREB: Nuclear Accumulation Assay

Once ERK1/2 is activated, a cascade of events is initiated, one chain of which results in activation of a set of kinases that includes p90 RSK and MSK kinases. These kinases phosphorylate the transcription factor CREB. The assay we describe here monitors the accumulation of phosphorylated CREB (on serine 133) in the nucleus. Using phosphorylation-specific primary antibodies and secondary antibodies coupled to the fluorochrome Alexa 647, only the phosphorylated form of CREB is detected/labeled (**Fig. 4B**). The nucleus is counterstained with Hoechst 33342. As readout, the ratio of intensity calculated from the phosphorylated CREB in the nucleus and in the cytoplasm of the cells was used. To analyze the images, we use the "Nuclear Traffic Module" of the INCell Analyzer 3000 software. The hardware and software settings are as described above.

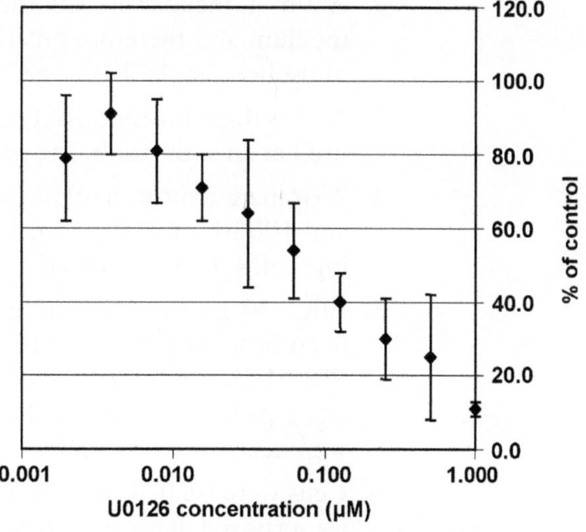

Fig. 3. TPA-induced phosphorylation of ERK dependent on MEK1. Cells were seeded as described above. After 16 h, increasing concentrations of the MEK1 inhibitor U0126 were added for 1 h. HUVECs were then stimulated with TPA for 30 min, and fixed and stained as indicated in the text. Phosphorylation and activation of ERK is expressed as nuclear intensity of phospho-ERK1/2 staining relative to staining in the absence of U0126 (% of control; 100% = 0 μM U0126), as a function of U0126 concentration.

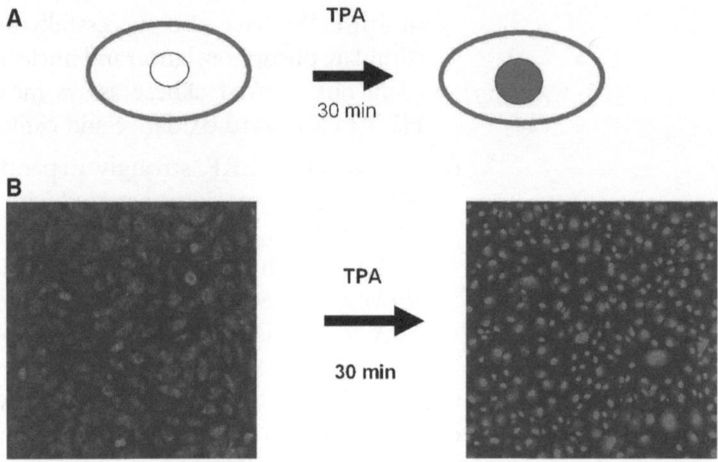

Fig. 4. Nuclear accumulation of phosphorylated CREB. (A) Phosphorylation and nuclear accumulation of phosphorylated CREB (Ser133) after TPA treatment. (B) HUVECs were stimulated with TPA as described in the main text. To detect distribution of transcription factor CREB, cells were fixed and permeabilized, and then initially incubated with a monoclonal antibody against the phosphorylated form (Ser133) of CREB and subsequently with secondary anti-mouse-IgG antibodies coupled to Alexa Fluor® 647. The left panel shows images of nonstimulated cells, and the right panel shows images of TPA-stimulated cells.

1. HUVECs were seeded at 10,000 cells per well (96-well Packard view plate, as described above) and incubated for 96 h in EGM2 medium at standard conditions.
2. A 96 h incubation will deplete growth factors from the medium and therefore greatly reduce the baseline amount of phosphorylated CREB (*see* **Note 5**).
3. As was done for the ERK1/2 assay, 1 μM TPA for 30min was used as an activator in the present assay (**Fig. 4**).
4. To initiate activation of the pathway, the medium was removed, and 100 μL of EGM-2 medium (not supplemented) containing 1 μM TPA was added.
5. After 30 min incubation at standard conditions, cells were fixed by adding 35 μL of 16% paraformaldehyde (*see* **Note 3**). **Fig. 4** shows images of HUVECs treated with (**Fig. 4B**, right panel) or without TPA (**Fig. 4B**, left panel). Images were taken with the INCell Analyzer 3000.
6. Cells were stained with the phosphorylation-specific antibodies against CREB as described above. As shown in **Fig. 5**, the nuclear accumulation of phosphorylated CREB shows a strong dependence on the concentration of TPA.
7. The plateau is reached at 250 nM of TPA and the AC_{50} is 15nM. TPA-induced nuclear accumulation of phosphorylated CREB is mediated by active MEK as shown in **Fig. 6**.

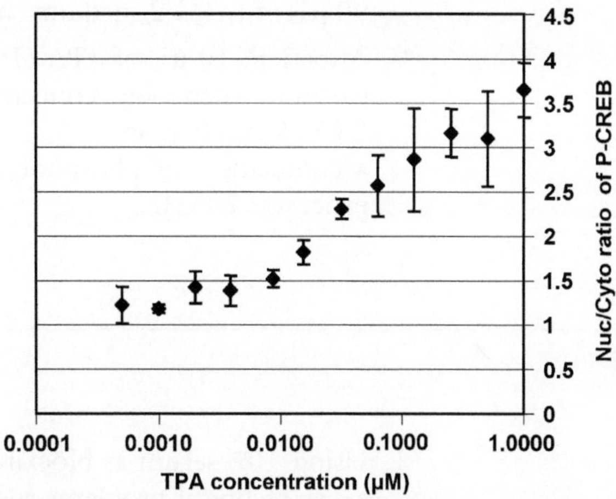

Fig. 5. TPA inducing phosphorylation of Ser133 of CREB in a dose-dependent manner. HUVECs were stimulated (96 h after seeding) for 30 min with the indicated concentrations of TPA. Cells were fixed and stained as described in Fig. 3. Activation is expressed as the nuclear-to-cytoplasmic ratio of the fluorescent signal as a function of TPA concentration.

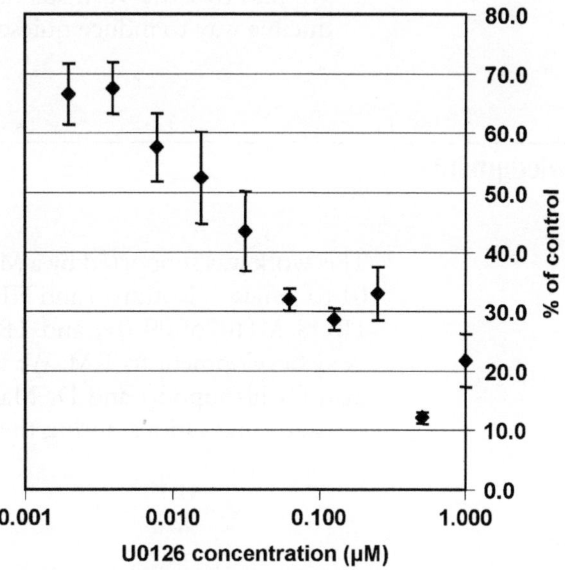

Fig. 6. Phosphorylation of CREB dependent on MEK1/ERK1/2 activation. Cells were seeded as described above. After 96 h, increasing concentrations of the MEK1 inhibitor U0126 were added for 1 h. HUVECs were then stimulated with TPA for 30 min and fixed and stained as indicated in the protocol section. Activation is expressed as the nuclear-to-cytoplasmic ratio of phospho-CREB staining relative to staining in the absence of U0126 (% of control; 100% = 0 μM U0126), as a function of U0126 concentration.

8. Ninety-six hours after seeding, cells were treated with increasing concentrations of U0126, a specific MEK1 inhibitor, in 90 μL of EGM-2 medium (not supplemented).
9. After 1 h, 10 μL of TPA (10 μM) was added to activate the pathway. Increasing concentrations of the MEK1 inhibitor U0126 resulted in a dose-dependent inhibition of nuclear accumulation of phosphorylated CREB. The IC_{50} for this process is 70nM.

4. Notes

1. Using 10% serum as blocking solution is much less likely to cause technical problems such as contamination or clogging than blocking reagents based on milk powder.
2. We routinely buy outdated frozen newborn calf serum to reduce costs.
3. In general, we find that it is not necessary to remove the medium or wash the cells before adding the fixative.
4. For most antibodies, 20-min incubation at room temperature is sufficient.
5. We find that the 96-h starvation period is an easy and reproducible way to induce quiescence in HUVECs.

Acknowledgments

This work was supported by a MLSCN grant #1U54 H6003914-01 to James E. Rothman and NIH Grants 1RO3 MH076344-01, 1RO3 MH076509-01, and 1RO3 MH076343-01, to support assay development to T.M. We would like to thank Dr J.E. Rothman for his support and Dr Martin Wiedmann for his input and constant suggestions during the progress of the work.

References

1. Robinson, M. J. and Cobb, M. H. (1997) Mitogen-activated protein kinase pathways. *Curr. Opin. Cell Biol.* 9,180–186.
2. Garrington, T. P. and Johnson, G. L. (1999) Organization and regulation of mitogen activated protein kinase signaling pathways. *Curr. Opin. Cell Biol.* 11, 211–218.
3. Chang, L. and Karin, M. (2001) Mammalian MAP kinase signalling cascades. *Nature* 410, 37–40.
4. Johnson, G. L. and Lapadat, R. (2002) Mitogen-activated protein kinase pathways mediated by ERK, JNK, and p38 protein kinases. *Science* 298, 1911–1912.

5. Sebolt-Leopold, J. S. and Herrera, R. (2004) Targeting the mitogen-activated protein kinase cascade to treat cancer. *Nat. Rev. Cancer* 4, 937–947.
6. Thompson, N. and Lyons, J. (2005) Recent progress in targeting the Raf/MEK/ ERK pathway with inhibitors in cancer drug discovery. *Curr. Opin. Pharmacol.* 5, 350–356.
7. Geller, H. M., Quinones-Jenab, V., Poltorak, M., and Freed, W. J. (1991) Applications of immortalized cells in basic and clinical neurology. *J. Cell. Biochem.* 45, 279–283.
8. Ulrich, A. B., Schmied, B. M., Standop, J., Schneider, M. B. and Pour, P. M. (2002) Pancreatic cell lines: A review. *Pancreas* 24, 111–120.
9. Torii, S., Nakayama, K., Yamamoto, T. and Nishida, E. (2004) Regulatory mechanisms and function of ERK MAP kinases. *J. Biochem. (Tokyo)* 136, 557–561.
10. Yoon, S. and Seger, R. (2006) The extracellular signal-regulated kinase: Multiple substrates regulate diverse cellular functions. *Growth Factors* 24, 21–44.
11. Gimbrone Jr, M. A., Cotran, R. S. and Folkman, J. (1974) Human vascular endothelial cells in culture: Growth and DNA synthesis. *J. Cell Biol.* 60, 673–684.
12. Mayer, T., Jagla, B., Wyler, M. R., Kelly, P. D., Aulner, N., Beard, M., Barger, G., Többen, U., Deborah, H., Smith, D. H., Brandén, L., and Rothman, J. E. (2006) Cell based assays using primary endothelial cells to study multiple steps in inflammation. *Meth. Enzymol.* 414, 266–283.
13. Naylor, L. H. (1999) Reporter gene technology: The future looks bright. *Biochem. Pharmacol.* 58, 749–757.
14. Chamberlain, C. and Hahn, K. M. (2000) Watching proteins in the wild: Fluorescence methods to study protein dynamics in living cells. *Traffic* 1, 755–762.
15. Favata, M. F., Horiuchi, K. Y., Manos, E. J., Daulerio, A. J., Stradley, D. A., Feeser, W. S., Van Dyk, D. E., Pitts, W. J., Earl, R. A., Hobbs, F., Copeland, R. A., Magolda, R. L., Scherle, P. A., and Trzaskos, J. M. (1998) Identification of a novel inhibitor of mitogen-activated protein kinase kinase. *J. Biol. Chem.* 273, 18623–18632.

Chapter 4

Large-Scale Small-Molecule Screen Using Zebrafish Embryos

Charles C. Hong

Summary

Zebrafish represents a versatile model organism with many molecular, morphological, and physiological similarities to mammals. Importantly, zebrafish are readily susceptible to perturbations by small molecules, including numerous pharmaceuticals in clinical use. Given these qualities, plus their small size and transparency, zebrafish embryos can be utilized for large-scale phenotype-based screens for small-molecule modifiers of biological processes. Thus, in a manner analogous to classical genetic screens, zebrafish chemical screens have the potential to reveal novel insights into complex biological pathways, as well as to identify lead compounds for novel therapeutics.

Key words: Chemical genetics, Chemical screen, Drug discovery, Large scale, Phenotype, Small-molecule library, Zebrafish.

1. Introduction

For centuries, man relied on astute, serendipitous observations for discovery of bioactive small molecules for medicinal purposes. In modern times, these chance discoveries have largely been supplanted by large-scale, high-throughput assays for chemical modifiers of purified proteins or cell extracts. Nevertheless, in principle, the phenotype-based approach to small-molecule discovery retains certain advantages. For example, compounds discovered to cause discernible effects on live animals are inherently closer to validation as drug leads than those discovered by target-based in vitro approaches *(1)*. Importantly, since the phenotype-based approach does not depend upon a priori selection

P.A. Clemons et al. (eds.), *Cell-Based Assays for High-Throughput Screening, Methods in Molecular Biology, Vol. 486*

DOI: 10.1007/978-1-60327-545-3_1

of a molecular target, it can be utilized to uncover novel or unanticipated biological insights *(2)*.

In the past few years, the zebrafish has emerged as an important model organism for small-molecule discovery *(1, 3, 4)*. Using zebrafish, it is now possible to combine the advantages of phenotype-based small-molecule discovery and modern high-throughput screening capabilities. From our experience, an important advantage of zebrafish chemical screening over other high-throughput platforms is the built-in means to assess specificity, efficacy, and toxicity of small molecules in the context of whole live animals (*see* **Note 1**). Another important aspect of chemical screens using zebrafish, in comparison to other model organisms such as *Drosophila* and *C. elegans*, is the close similarity between zebrafish and mammalian orthologs. Thus, small molecules discovered in zebrafish screens typically have directly analogous effects on mammalian systems, and vice versa *(2, 5–8)*.

Large-scale chemical screens in zebrafish have successfully identified small molecules that modulate numerous developmental

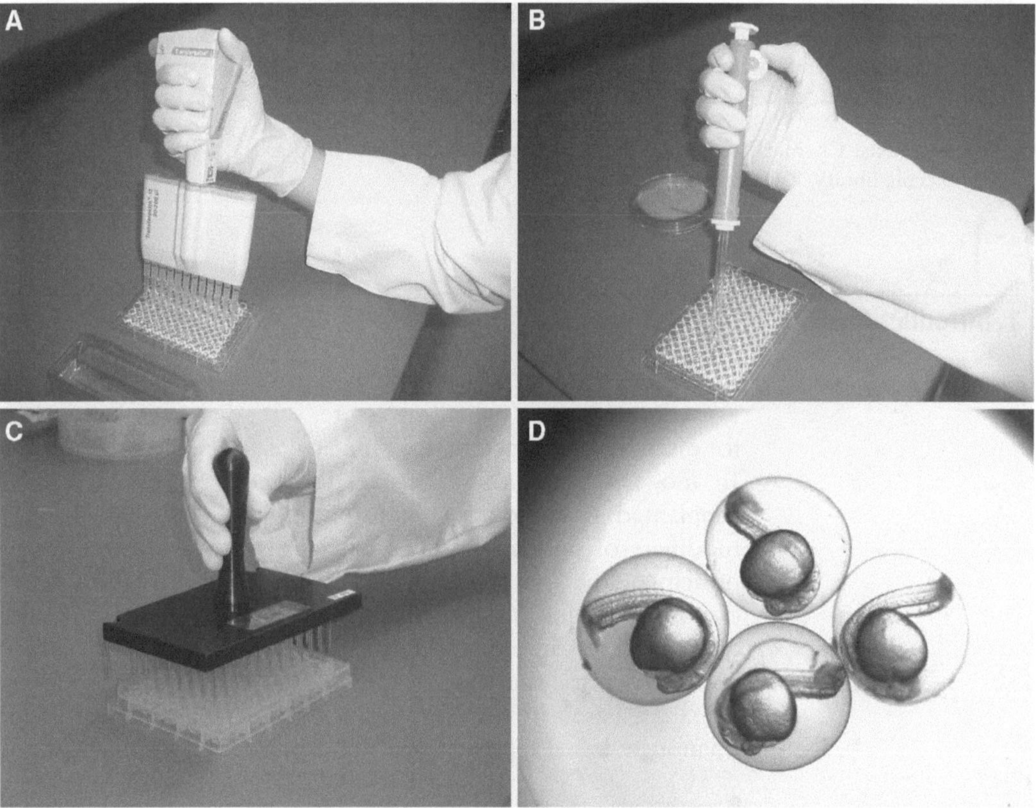

Fig. 1. Basic methodology for chemical screening using zebrafish embryos. (A) Using a multichannel pipettor, 100 μL of E3 medium is transferred to each well of a 96-well assay plate (recipient plate). (B) Using a glass Pasteur pipette and manual pipette pump, embryos are transferred to the recipient plate. (C) Pin-transfer of compounds from the source plate to the recipient plate. (D) Direct visual examination of embryo phenotype using a stereomicroscope.

processes as well as disease models with important therapeutic implications *(2, 5, 9, 10)*. In addition, borrowing from the logic of classical genetic modifier screens, zebrafish have been successfully utilized to discover unanticipated roles of the phosphatidylinositol-3 kinase (PI3K) and the Erk/MAP kinase signaling pathways during artery/vein specification *(2)*. Moreover, the first selective small-molecule BMP inhibitor, which holds significant investigational and therapeutic potential, was recently identified in a phenotype-based screen for compounds that perturb the zebrafish dorso-ventral axis (*13, 14*).

While zebrafish chemical screens can be modified to attain high-throughput capabilities through robotic small-molecule transfer technologies, automated sorting, and phenotyping, the basic methodology is very straightforward: embryos are arrayed in multiwell plates, compounds added by either pin-transfer or by high-throughput liquid transfer methods, and then embryonic phenotypes caused by compounds analyzed (**Fig. 1**). This chapter focuses on basic reagents and methods available to most academic investigators to perform a phenotype-based chemical screening using zebrafish embryos. It is hoped that large-scale zebrafish chemical screens, which require a fertile imagination but relatively modest capital investment, will be broadly adopted for discovery of numerous chemical tools to study biology and for development of new therapies to treat human diseases.

2. Materials

One of the most important decisions in planning a zebrafish chemical screen involves the selection of the appropriate fish strain and disease model. One must, for example, consider the potential for informative and meaningful dissection of a particular developmental pathway and/or its general relevance as a disease model. In addition, when planning a screen for chemical suppressors of a mutant or diseased phenotype, penetrance and robustness of the phenotype must be thoroughly evaluated (*see* **Note 2**). We have successfully utilized homozygous viable *gridlock* (*grl*) mutant embryos in our large-scale chemical screens *(2, 5)*. An advantage of a homozygous, viable mutant strain is that, when mutant females are mated to mutant males, all of the progeny will be mutant. In the case of *grl*, the mutant phenotype typically has greater than 95% penetrance. Thus, when three embryos are arrayed into each well, the probability of all three having a phenotypic suppression (wild-type phenotype in mutant embryos) by chance alone is only 1 in 8,000 (0.0125%). Moreover, *gridlock* mutation causes a distinct loss of tail circulation that can be easily scored using a simple stereomicroscope *(11)*. In contrast, screening for small-molecule suppressors of a homozygous recessive mutation is more challenging, since 75% of embryos from heterozygote crosses will be phenotypically normal at baseline. Zon and colleagues overcame this challenge by placing 20 embryos per well, a format in which only 0.3% of the wells are predicted to contain all embryos with wild-type phenotype *(10)*.

The choice of chemical libraries to screen is another critical matter. Thanks in large part to recent advances in synthetic organic chemistry, one can gain access to hundreds of thousands of small molecules from numerous commercial sources, as well as from the Small Molecule Repository (SMR) of the Molecular Libraries Screening Centers Network (MLSCN; *see* **Note 3**). A major disadvantage of commercial libraries, in comparison to the MLSCN's libraries, is the cost; however, an important advantage of acquiring a library of one's own is that the design and conduct of zebrafish screens can be catered to the researcher's particular interest and expertise (*see* **Notes 4** and **5**).

It is important to remember that zebrafish chemical screening is not a very high-throughput platform. In contrast to molecule-based or cell-based chemical screens, which can screen 100,000 compounds rapidly in an automated fashion, zebrafish screens are limited by the number of eggs laid and by the laborious transfer of embryos into 96-well plates. In our experience, while up to 1,000 compounds can be screened in a given day, one can reasonably expect to screen only about 10,000 compounds in a month. To overcome this limitation, one can modify the screen such that each well contains a pool of compounds *(10)*. A disadvantage of this approach is the potential of dismissing real "hits" due to confounding effects of multiple small molecules (*see* **Note 6**). The last decision involves whether to screen a library of unknown synthetic compounds or of "known bioactives," typically comprising bioactive components of natural products and modern medicines. While known bioactive libraries are typically much smaller than commercial libraries of unknown synthetic compounds, an underappreciated aspect of known bioactives is that they typically have significantly higher "hit" rates than libraries of synthetic compounds, and that "off-target" effects of known bioactives sometimes can be very informative. Thus, given the inherent limitations of a zebrafish chemical screen, it seems prudent to begin with a library of known bioactive molecules (*see* **Note 7**).

2.1. Zebrafish Egg Collection

1. Minimum of 20 pairs of adult zebrafish of desired genotype.
2. Fish nets (Aquatic Habitats; Apopka, FL).
3. Breeding tanks (1–2 L capacity), with removable inner container and dividers (Aquatic Habitats).

4. Petri dishes (10 cm) (Fisher Scientific; Pittsburgh, PA).
5. Plastic tea strainer (Fackelmann; Germany).
6. Wash bottle containing embryo water (Nalgene; Rochester, NY).
7. Disposable polyethylene transfer pipettes (Fisher Scientific).

2.2. Sorting of Embryos into Multi-well Plates

1. Polystyrene 96-well round-bottom assay plates (Corning COSTAR; Lowell, MA).
2. Glass Pasteur pipette (9 in.) (Fisher Scientific).
3. Manual pipette pump, 10 mL (Bel-Art Products; Pequannock, NJ).
4. E3 embryo medium: 5 mM NaCl, 0.17 mM KCl, 0.33 mM $CaCl_2$, 0.33 mM $MgSO_4$, containing 0.003% PTU (phenylthio-carbamide, Sigma; St. Louis, MO). PTU can be prepared as a 10× solution by dissolving 0.3-g PTU in 1 L of E3 embryo media. Solutions containing PTU should be protected from light by covering with aluminum foil.
5. Pipettor :12-channel, 20–200 μL (BrandTech; Essex, CT).
6. Disposable polystyrene pipette basin, 50 mL (Fisher Scientific).

2.3. Small-Molecule Library

1. Small-molecule library of structurally diverse compounds arrayed in a 96-well format at 10 mM stock in DMSO. In recent screens, we have used DiverSet E library from Chembridge (*see* **Notes 4** and 7). Each master plate is aliquoted into several 96-well polypropylene storage plates (Corning), and stored at –80°C until use.
2. Aluminum sealing tape for 96-well plates (Nunc; Rochester, NY).
3. Multiblot replicator, 96-pin, approximately 100 nL/drop (V & P Scientific; San Diego, CA).
4. Dessication chamber (Fisher Scientific) containing Drierite (anhydrous calcium sulfate, W.A. Hammond Drierite Co; Xenia, OH).
5. Tabletop centrifuge with 96-well plate adaptors (Eppendorf; Westbury, NY).
6. Plastic cover for a micropipette tip rack.
7. Bunsen burner.
8. DMSO (Sigma; St. Louis, MO).
9. Ethanol.

2.4. Screening by Visual Inspection

1. Basic incubator, 28.5°C (Fisher Scientific).
2. Stereomicroscope with transmitted light base (Leica Microsystems; Bannockburn, IL) or compound inverted microscope for tissue culture (Fisher Scientific).

3. Methods

In classical forward-genetic screens, a single gene can cause myriad phenotypes depending on nature of the mutation. A similar concept holds true for chemical screens. A small molecule found to perturb one particular aspect of the zebrafish embryo when added at one specific dose and time in embryogenesis might have very different effects when added at different doses or times. Thus, by varying the doses of compounds and the timing of compound addition, one can perform several distinct screens using the same small molecule library.

Borrowing further from the analogy to forward-genetic screens, the adage "you find what you seek" applies for chemical screens as well. For example, if one desires to identify small molecules that have vascular effects but do not perturb other aspects of development, one can add chemicals early in development, perhaps at the sphere stage. A compound with strictly vascular-selective effects would be predicted not to interfere with any of the early developmental decisions, whereas less specific compounds might cause gross abnormalities or early embryonic lethality when added at such an early stage. Conversely, if one desires to identify small molecules that have pleiotropic effects, one can add chemicals at the precise time when a particular developmental decision is being made *(2)*.

The main advantage of zebrafish chemical screens over higher-throughput cell- and molecule-based platforms is that in vivo efficacy and specificity can be quickly and directly assessed in a whole zebrafish embryo (*see* **Notes 1** and **8**). By screening at the relatively low dose of 1–2.5 μM, one can enrich for discovery of potent compounds. Such compounds, however, may have significant off-target effects at higher doses. In contrast, at the relative high dose of 20–50 μM, one enriches for compounds with greater specificity (*see* **Note 9**). Lastly, it is important to recognize that once a larva's mouth and gills become functional at 5 days post fertilization, sensitivity to compounds drops substantially, even by 1 order of magnitude [*(6)*, personal observations].

As described in the preceding and subsequent chapters, the screening process can be automated. Since the zebrafish embryo is not an exceptionally high-throughput platform in the first place, the main advantage of adding automation to the screening process lies not in substantially increasing throughput, but in enabling examination of the embryos at unusual hours or at frequent intervals. As in forward-genetic screens, observers trained to distinguish meaningful phenotypic changes remain essential to chemical screens in zebrafish. Also, as in forward-genetic screens, phenotypic analysis can be greatly assisted by the use of transgenic strains expressing fluorescent proteins in an organ-specific manner *(12)*.

3.1. Collection of Embryos

1. Mating crosses are set up in the afternoon prior to the day planned for egg collection. Ten to twenty breeding tanks are filled with water from the aquaculture system. Using the fish net, adult fish are transferred to the breeding tanks, each tank containing one male and one to two females in the inner container. Male and female fish will be separated from each other by a divider. Label the cages and put a lid over them to prevent fish from jumping out.
2. In the morning, the dividers are taken out of the breeding tanks and zebrafish allowed to mate and spawn. Over the next 1–2 h, eggs are laid and fall through a grid at the bottom of the inner container and collect in the outer container, where the adults cannot eat them.
3. After 1–2 h, examine each breeding tank for eggs collected at the bottom following a successful mating. Return adult fish back to permanent storage tanks using the fish net. Remove the inner container, and strain the water in each breeding tank through a plastic tea strainer. Eggs will collect in the strainer. Invert the strainer over a Petri dish and, using a wash bottle containing the E3 medium, gently rinse the strainer to flush the eggs into the Petri dish. Average mating success can range from 50 to 100%, and a single mating yields between 100 and 400 eggs. As soon as possible, all unfertilized eggs, which look cloudy white, should be removed using a disposable plastic pipette. Repeat this step as necessary.

3.2. Arraying of Embryos in 96-Well Plates

Transfer of embryos to a 96-well plate is the most labor-intensive step in zebrafish chemical screens (*see* **Note 10**). Embryo transfer can be automated using a COPAS XL automated sorter (Union Biometrica; Holliston, MA) but as this requires a substantial investment beyond the means of most laboratories, a manual method is described.

1. Pour about 50 mL of E3 medium containing PTU into a disposable pipette basin (*see* **Note 11**). Using a 12-channel pipettor, transfer 100 μL of E3 medium from pipette basin to each well of a round-bottom 96-well assay plate.
2. Prepare a glass Pasteur pipette by cutting the tip with a glass-cutter to widen the opening and polishing the tip with a flame from a Bunsen burner. Connect the glass pipette to the manual pipette pump.
3. To draw up fluid and eggs into the glass pipette, submerge the tip of the pipette into the E3 medium containing the eggs, and turn the operating wheel of the pipette pump with the thumb. On average, about 20–50 eggs are drawn up each time.
4. Hold the pipette pump upright and agitate fluid collected inside the glass pipette by gently tapping or moving the

plunger up and down with the operating wheel. This will allow eggs to settle near the tip of the glass pipette.

5. Turn the operating wheel of the pipette pump slightly to create a mild positive pressure within the pipette but not enough to expel the eggs. While exerting a positive pressure, touch the pipette tip to the meniscus of E3 medium in each well, and quickly withdraw the pipette. This will draw by capillary action a tiny amount of fluid from the glass pipette to each well, carrying 1–3 eggs along with it. This procedure minimizes the total amount of E3 medium transferred along with the eggs. Repeat this motion as necessary to transfer three embryos to each well. With practice, embryos can be transferred to several 96-well plates in 1–2 h. Alternatively, three embryos can be transferred to each well by actively expelling them. This will result in transfer of variable amounts of medium to each well. To ensure that each well contains roughly the same amount of liquid, E3 medium can first be removed using a multichannel pipette, leaving just the embryos, and then 100 μL of E3 medium put back into each well.

3.3. Transfer of Small-Molecule Library

A typical small molecule library is supplied in a 96-well format, with each compound stored in DMSO as a 10 mM stock (*see* **Notes 4** and 7). While compound transfer can be automated with robotic transfer methods, this chapter will describe methods using manual pin transfer.

1. Thaw at room temperature the desired number of 96-well storage plates containing aliquots of small molecule library (source plates). Take note of the serial or other identification number of the source plates used. To minimize condensation outside the plates, thawing can occur in a desiccation chamber containing Drierite.
2. Gently mix the source plates on an Eppendorf Mix Mate at 300 rpm or by gently swirling to resuspend or dissolve any compounds that might have precipitated out during storage (*see* **Note 12**).
3. Briefly (a few seconds) spin down the plates in a tabletop centrifuge equipped with multiwell plate adaptor.
4. Remove the aluminum sealing tape from source plate.
5. Carefully align the 96-pin multiblot replicator over the 96 wells on the source plate. Slowly lower the replicator and dip the replicator's pins into wells of the source plate. Gently mix contents of wells with the replicator, and then remove the pins from the wells at a slow and even speed (~0.5 cm/s). Removing the pins from the source plate quickly produces larger hanging drops and hence greater amount of liquid transferred. Deliver the compounds on the replicator's pins

to 96-well plate containing embryos and E3 media (recipient plate) by dipping and raising the pins three times through the recipient plate's meniscus. Take care to ensure that the recipient plate is correctly aligned. With a VP409 multiblot replicator, this technique transfers approximately 100 nL of compound to each well. The final concentration of each compound is 10 μM (*see* **Note 13**).

6. Record the identification number of the source plate on the recipient plate.
7. Clean the multiblot replicator by rinsing the pins with DMSO inside a small basin (inverted lid of a pipette tip rack), followed by two rinses in distilled water. Dry the pins by dipping them in ethanol inside an inverted tip lid box, and briefly flaming the alcohol off using a Bunsen burner. Take care not to heat the pins directly in a Bunsen burner, as this will damage the replicator.
8. Repeat **steps 4–6** for each source plate as needed.
9. Cover the recipient plate containing the embryos and compounds with the appropriate lid, and place in the 28.5°C incubator.
10. Cover each source plate containing small molecule library with an aluminum sealing tape. Ensure that the tape forms a tight seal using a rolling sealer. Put the source plates back in a –80°C freezer for long-term storage.

3.4. Screening for Effects of Small Molecules by Visual Inspection of Phenotypes

1. Prior to performing the screen, formulate a specific criterion for what would constitute a "hit."
2. At desired times in development, remove the 96-well plates containing compound-treated embryos from the incubator and examine each well under a stereomicroscope. For better visualization of subtle changes, such as changes in circulatory pattern, a phase-contrast inverted microscope can be used. Fluorescence microscopy can be used to examine perturbation of expression of GFP or DsRed proteins under a tissue-specific promoter. Various methods to automate the screening process are discussed in subsequent chapters.
3. Quickly scan the 96-well plates for any well in which at least two out of three embryos exhibit the prescribed "hit" phenotype. Record the identity of the plate and the well location of each potential "hit." In addition, note the wells containing embryos with any interesting phenotypes even if they are not necessarily a "hit."

3.5. Confirmation of a Potential "Hit"

1. Retrieve the compound library (source plate) from which a potential "hit" was found. Thaw, mix, and spin down the source plate as described above.

2. Remove the aluminum sealing tape, and locate the well containing the compound that produced the "hit" phenotype.
3. Remove a small aliquot (10 μL) of the potential "hit" from the source plate.
4. Using this aliquot, retest the effects of the compound at several doses (1, 5, 10, and 50 μM). For each dose, ten embryos are tested in 0.5 mL of E3 media in a 48-well plate format (*see* **Note 14**). The timing of compound addition for retesting should be identical to that of the original screening.
5. A "hit" is confirmed when the elicited phenotype is reproduced upon retesting of the compound (*see* **Note 14**).
6. Identity of the "hit" compound is determined from the database of small molecules in the chemical library.

4. Notes

1. In addition to large-scale chemical screens, whole zebrafish embryos can be utilized for structure–activity relationship (SAR) studies to rapidly identify chemicals with improved potency and efficacy without increased toxicity *(1, 10)*.
2. When considering a screen for chemical suppressors of an inducible phenotype, particular attention must be paid to the background and induced frequencies. In the case of a phenotype caused by a mutant gene expressed under heat-shock control, the condition that induces the phenotype with greatest reproducibility must be carefully mapped out prior to initiating a screen to avoid unacceptably high false-positive rates.
3. The Molecular Libraries Screening Center Network (MLSCN) was established as a part of the NIH Roadmap. MLSCN's Molecular Libraries Small-Molecule Repository (MLSMR) will eventually consist of 500,000 natural products, known bioactives, and a large-diversity set of compounds. While access to MLSCN's compounds is free, screens themselves are performed at 1 of the 10 designated centers. Of these, the Universities of Pennsylvania and Pittsburgh currently have the capability to perform zebrafish screening. Importantly, compounds identified as "hits" from screens done through the MLSCN are considered pre-competitive, and the MLSCN requires that screening results be promptly deposited in PubChem (http://pubchem.ncbi.nlm.nih.gov). Latest information, funding mechanisms, and specific guidelines on submitting an assay method to a particular center within the MLSCN can be obtained from its website

(http://mli.nih.gov/mlscn/). In addition, there are numerous university-based screening programs that provide access to small molecule libraries. For example, The Broad Institute of Harvard and MIT's Chemical Biology Program is open to the larger public research community (http://www.broad.harvard.edu/chembio). Finally, *ChemBank* (http://chembank.broad.harvard.edu) and PubChem (http://pubchem.ncbi.nlm.nih.gov) are two major small-molecule databases open to the research community.

4. Listed here are some commercial vendors that provide small-molecule libraries (this list is not meant to be comprehensive, as the number of vendors and libraries is expanding rapidly): Biomol (http://www.biomol.com), Chembridge (http://www.chembridge.com), ChemDiv (http://www.chemdiv.com), Maybridge (http://www.maybridge.com), Microsource (http://www.msdiscovery.com), Sigma (http://www.sigmaaldrich.com), and TimTec (http://www.timtec.net).

5. Web portals provided by numerous commercial vendors can be used to quickly obtain structural analogs of a primary "hit" for preliminary SAR studies.

6. Only about 2% of small molecules in a typical commercial library cause death or severe developmental delay of zebrafish embryos *(10)*. Thus, if five compounds are pooled per well, death will occur in about 10% of wells. In such a scenario, compounds causing death when pooled can be retested individually at a later time.

7. Libraries of known bioactive small molecules are available from many commercial sources, including Biomol's ICCB Known Bioactives 1 and 2 (480 compounds each), Micro-Source Discovery System's NINDS Custom Collection 2 (1,040 compounds) and Spectrum Collection (2,000 compounds), Prestwick's Collection (1,120 of out-of-patent drugs), and Sigma's LOPAC (1,280 pharmacologically active compounds) Collection.

8. For biochemical or cell-based assays, initial "hit" rates can be high. For example, a screen for activators of deacetylase activity of purified Sirt1 found 1,151 initial "hits" out of 2,139 molecules screened (*ChemBank*). For such cases, a number of reiterations of assays as well as alternative tests are necessary to validate initial "hits." By contrast, a screen of over 5,000 compounds found only one that specifically perturbed embryonic dorso-ventral axis (Hong and Peterson, manuscript in preparation).

9. It is interesting to note that 20 and 50 μM are relatively moderate concentrations, and several very important chemical probes, such as cyclopamine, might not have passed muster under such screening conditions.

10. Transfer of hatched larvae can be particularly challenging given their mobility. To screen for small molecules affecting a larval phenotype, one option might to be to transfer embryos to 96-well plates prior to hatching, and add the small molecules at a later time.
11. E3 medium with PTU can be supplemented with antibiotics (2 μM metronidazole, 5 units/mL penicillin, 50 μg/mL streptomycin; diluted from a 100× stock solution) to prevent bacterial and fungal growth *(10)*.
12. After long-term storage at –80°C, some of the compounds in the source library plate may precipitate out of DMSO solution. If some compounds remain out of solution even after 15-min mixing, it may be prudent to proceed with the screen anyway.
13. Final compound concentrations can be changed by varying the amount of E3 in the recipient well or repeating the pin transfer. It is inevitable that some compounds will precipitate out of solution in E3 medium.
14. It is not unusual for a putative "hit" to cause the induced phenotype at a different concentration than that of the screen itself. Moreover, retesting may not reproduce the phenotype in 100% of tested embryos. Multiple retesting, perhaps done blinded, may be necessary to reach a statistically significant difference between treated and untreated embryos. A careful retesting of each "hit" at this early stage can prevent wasting valuable time and effort on false positives.

Acknowledgments

The author would like to thank Professor Randall T. Peterson for his advice and encouragement. This work was supported by Sarnoff Endowment for Cardiovascular Research and K08 HL081535.

References

1. Zon, L. I. and Peterson, R. T. (2005) In vivo drug discovery in the zebrafish. *Nat. Rev. Drug Discov.* 4, 35–44.
2. Hong, C. C., Peterson, Q. P., Hong, J. Y., and Peterson, R. T. (2006) Artery/vein specification is governed by opposing phosphatidylinositol-3 kinase and MAP kinase/ERK signaling. *Curr. Biol.* 16, 1366–1372.
3. MacRae, C. A. and Peterson, R. T. (2003) Zebrafish-based small molecule discovery. *Chem. Biol.* 10, 901–908.
4. Peterson, R. T., Link, B. A., Dowling, J. E., and Schreiber, S. L. (2000) Small mol-

ecule developmental screens reveal the logic and timing of vertebrate development. *Proc. Natl. Acad. Sci. USA* 97, 12965–12969.

5. Peterson, R. T., Shaw, S. Y., Peterson, T. A., Milan, D. J., Zhong, T. P., Schreiber, S. L., et al. (2004) Chemical suppression of a genetic mutation in a zebrafish model of aortic coarctation. *Nat. Biotechnol.* 22, 595–599.
6. Bayliss, P. E., Bellavance, K. L., Whitehead, G. G., Abrams, J. M., Aegerter, S., Robbins, H. S., et al. (2006) Chemical modulation of receptor signaling inhibits regenerative angiogenesis in adult zebrafish. *Nat. Chem. Biol.* 2, 265–273.
7. Carmeliet, P. (2003) Angiogenesis in health and disease. *Nat. Med.* 9, 653–660.
8. Milan, D. J., Peterson, T. A., Ruskin, J. N., Peterson, R. T., and MacRae, C. A. (2003) Drugs that induce repolarization abnormalities cause bradycardia in zebrafish. *Circulation* 107, 1355–1358.
9. Lawson, N. D., Vogel, A. M., and Weinstein, B. M. (2002) Sonic hedgehog and vascular endothelial growth factor act upstream of the Notch pathway during arterial endothelial differentiation. *Dev. Cell* 3, 127–136.
10. Stern, H. M., Murphey, R. D., Shepard, J. L., Amatruda, J. F., Straub, C. T., Pfaff, K. L., et al. (2005) Small molecules that delay S phase suppress a zebrafish bmyb mutant. *Nat. Chem. Biol.* 1, 366–370.
11. Weinstein, B. M., Stemple, D. L., Driever, W., and Fishman, M. C. (1995) Gridlock, a localized heritable vascular patterning defect in the zebrafish. *Nat. Med.* 1, 1143–1147.
12. Burns, C. G., Milan, D. J., Grande, E. J., Rottbauer, W., MacRae, C. A., and Fishman, M. C. (2005) High-throughput assay for small molecules that modulate zebrafish embryonic heart rate. *Nat. Chem. Biol.* 1, 263–264.
13. Yu, P. M.*, Hong, C. C.*, Sachidanandam, C.*, Babitt, J. L., Deng, D. Y., Hoyng, S. A., Lin, H. Y., Bloch, K. D., Peterson, R. T. (2008) Dorsomorphin inhibits BMP signals required for embryogenesis and iron metabolism. Nat. Chem. Biol. 4:33–61. *equal contribution.
14. Hao, J. J., Daleo, M. A., Yu., P. M., Murphy, C. K., Ho, J. N., Hu, J. N., Hu, J., Peterson, R. T., Hatzopoulos, A. K., Hong, C. C. (2008) Dorsomorphin, a selective small molecule inhibitor of the BMP signaling, promotes cardiomyogenesis in embryonic stem cells. *PLoS ONE* 3, e2904.

Chapter 5

Whole-Animal High-Throughput Screens: The *C. elegans* Model

Eyleen J. O'Rourke, Annie L. Conery, and Terence I. Moy

Summary

The nematode *Caenorhabditis elegans* shows a high degree of conservation of molecular pathways related to human disease, yet is only 1-mm long and can be considered as a microorganism. Because of the development of a simple but systematic RNA-interference (RNAi) methodology, *C. elegans* is the only metazoan in which the impact of "knocking-down" nearly every gene in the genome can be analyzed in a whole living animal. Both functional genomic studies and chemical screens can be carried out using *C. elegans* in vivo screens in a context that preserves intact cell-to-cell communication, neuroendocrine signaling, and every aspect of the animal's metabolism necessary to survive and reproduce in lab conditions. This feature enables studies that are impossible to undertake in cell-culture-based screens. Although genome-wide RNAi screens and limited small-molecule screens have been successfully performed in *C. elegans*, they are typically extremely labor-intensive. Furthermore, technical limitations have precluded quantitative measurements and time-resolved analyses.

In this chapter, we provide detailed protocols to carry out automated high-throughput whole-animal RNAi and chemical screens. We describe methods to perform screens in solid and liquid media, in 96 and 384-well format, respectively. We describe the use of automated handling, sorting, and microscopy of worms. Finally, we give information about worm-adapted image analysis tools to quantify phenotypes. The technology presented here facilitates large-scale *C. elegans* genetic and chemical screens and it is expected to help shed light on relevant biological areas.

Key words: 384-well plate, 96-well plate, Agar, Antimicrobial, Automation, *C. elegans*, Chemical, Fluorescent marker, High-throughput, In vivo, Quantitative, RNAi, Screen, Whole-animal.

1. Introduction

Caenorhabditis elegans provides an ideal compromise between complexity and tractability. Many biological processes are conserved between humans and *C. elegans* to such an extent that

P.A. Clemons et al. (eds.), *Cell-Based Assays for High-Throughput Screening, Methods in Molecular Biology, Vol. 486*

DOI: 10.1007/978-1-60327-545-3_1

C. elegans data in many cases predict gene function, drug–target interaction, and target validation in mammals *(1–3)*. *C. elegans* investigations have already fostered a better understanding of the underlying mechanisms of a number of human diseases *(4–6)*. Moreover, although *C. elegans* is a sophisticated multicellular animal, all 959 somatic cells of its transparent body are visible with a microscope. Its constant body pattern, in combination with fluorescent markers, makes almost any aspect of *C. elegans* physiology suitable for in vivo study. Furthermore, an emerging set of informatic imaging technologies, such as WormTracker, allows automated behavioral analysis *(7)*.

In 2003, *C. elegans* became the first multicellular organism for which systematic RNAi experiments could be performed in a living animal *(8)*. siRNA can be delivered systemically simply by feeding *C. elegans* an *E. coli* strain expressing a double-stranded RNA (dsRNA) targeting a gene of interest *(9)*. The availability of a genome-wide RNAi feeding library has facilitated the completion of multiple genome-wide loss-of-function phenotypic screens *(10–12)*. Furthermore, by comparing the effects of RNAi in wild-type and mutant backgrounds, *C. elegans* allows large-scale analysis of functional genetic interactions *(13)*. Finally, the conservation of disease pathways and the cost-effectiveness of its cultivation make *C. elegans* an incomparable tool to discover new bioactive compounds and their targets. The combination of traditional genetics, RNAi technology, and small-molecule screens can help triangulate on potential drug targets, a major issue with any drug screen *(14)*.

Despite the potential value of *C. elegans* for gene, drug, and drug-target discovery, current screens are carried out manually, are extremely labor-intensive, use volumes incompatible with high-throughput chemical screens, and are not quantitative.

Here we present two protocols for automated and quantitative high-throughput (HT) *C. elegans* screening. The described methods enable quantitative analyses of a wide range of biological processes, such as the response to different types of biotic (pathogens) or abiotic (heavy metals, ultraviolet radiation, heat) stresses that affect viability, as well as traditional longevity studies. In addition, automated examination of any phenotypic readout based on fluorescent markers such as green fluorescent protein (GFP), Nile Red, or MitoTracker becomes feasible. Finally, the study of chemical or genetic perturbations that affect growth rates or body size could also be automated using the described methodologies.

To improve the throughput of RNAi screens, we developed a protocol to miniaturize the standard *C. elegans* RNAi screening method that is typically carried out in 6 or 12-well plates. In the presented HT assay, worms are grown in 96-well plates containing solid media, which prevents the metabolic shift manifested as slow growth and a starved aspect that occurs when *C. elegans* is grown or transferred to a liquid medium. Briefly, RNAi-feeding bacteria are grown overnight and then seeded in 96-well microtiter plates containing low-fluorescent agar media with IPTG to induce the synthesis of double-stranded RNA. In two-generation screens, two synchronized first larval stage (L1) worms, obtained by bleaching adults and allowing the embryos to hatch overnight without food, are dispensed into each well using a worm-handling robot. The plates are incubated for six days, allowing sufficient time for the two L1 worms to grow to adults, to lay eggs, and to allow the eggs to hatch and develop. In one-generation screens, twenty synchronized L1s are dispensed in each well and are incubated for three days to allow the 20 animals to reach adulthood. Images of wells are taken with an automated microscope and then analyzed with adapted image-analysis tools.

To enable chemical screens, we developed a method to screen for small molecules in 384-well plates. The described assay was designed to screen low-molecular weight compounds that prevent the lethal effect of infection by the bacterial pathogen *Enterococcus faecalis (15). C. elegans* are first grown to the young adult stage and then infected on the lawns of *E. faecalis.* The infected worms are washed and transferred to 384-well plates containing liquid media and the compounds to be tested. The plates are incubated until the infection kills untreated worms. The worms are washed and stained with a fluorescent dye that specifically stains dead worms. Images of the wells are captured with an automated microscope and analyzed to quantify worm survival.

1.1. Initial General Considerations Regarding RNAi and Chemical Screens

There are a number of characteristics that make an assay suitable for HT approaches involving *C. elegans.* The assay must be robust, reproducible (high Z`-score; *see* **Chap. 1**), and have a readout that is amenable to automated analysis. In addition, it must be possible to miniaturize the assay to a 96-well or higher-density format. Many variables need to be optimized to develop an assay, including genetic background, readout, food source, temperature, timing, number of animals needed to have sufficient statistical power, number of replicates, *etc.* All those variables will need to be adjusted to the 96 or 384-well format, where small perturbations such as salt concentration, amount of food, and number of animals will have a larger effect than in a standard assay, given the limitations that the small format imposes. An important feature of HT *C. elegans* assays is that sterile working conditions need to be maintained at all times, given that growth and handling of worms is carried out in pools and therefore contamination introduced at one step can ruin numerous assay plates.

2. Materials

2.1. RNAi Screens

2.1.1. Reagents

1. *C. elegans* strain NL2099 (*rrf-3* (*pk1426*)) *(16)*. Worms can be obtained from the *C. elegans* Genetics Center (http://www.cbs.umn.edu/CGC/Strains/request.htm) (*see* **Note 1**).
2. RNAi feeding library constructed in the Ahringer laboratory is available from Geneservice, Ltd. (http://www.geneservice.co.uk/products/rnai/index.jsp).
3. Potassium phosphate buffer, pH 6.0: 108.3 g KH_2PO_4, 35.6 g K_2HPO_4, water to 1 L. Sterilize by autoclaving.
4. NGM agar: 3 g NaCl, 2.5 g peptone, 17 g agar, 975 mL water. Autoclave. Cool to 55°C. Add in order the following sterile solutions: 1 mL of 5-mg/mL cholesterol dissolved in ethanol, 1 mL of 1 M $CaCl_2$, 1 mL of 1 M $MgSO_4$, and 25 mL of 1 M potassium phosphate, pH 6.0.
5. LB broth: 10 g Bacto-tryptone, 5 g yeast extract, 10 g NaCl, water to 1 L pH to 7.5 with NaOH. Sterilize by autoclaving. In some experiments LB is supplemented with 100 µg/mL of carbenicillin (LB-Cb).
6. LB-Amp/Tet plates: add 16 g/L of agar to LB-broth. Sterilize by autoclaving. After cooling, supplement with 100 µg/mL ampicillin and 25 µg/mL of tetracycline, and pour into Omnitray single-well square plates. Store at 4°C.
7. S-basal without cholesterol buffer (S-buffer): 5.85 g NaCl, 1 g K_2HPO_4, 6 g KH_2PO_4, water to 1 L. Sterilize by autoclaving.
8. *E. coli* OP50 bacteria: grow overnight in LB-broth, 200 rpm, at 37°C. Pellet bacteria, and concentrate 10 times in S-buffer (e.g., grow OP50 in 500 mL of LB-broth and resuspend in 50 mL of S-buffer). Store at 4°C. When needed seed 1 mL of OP50 suspension per 90 mm NGM agar plate and let dry.
9. Lysis solution: 195 µL 10N NaOH, 600 µL sodium hypochlorite (Aldrich; St. Louis, MO) or commercially available bleach brands, water to 3 mL.
10. Low-fluorescent RNAi media (LFR-media): 2 g NaCl, 2 g bactopeptone, 16 g agarose, water to 1 L. Sterilize by autoclaving with a magnetic bar per flask. After autoclaving, add in order the following sterile solutions: 1 mL of 5-mg/mL cholesterol dissolved in ethanol, 1 mL of 1 M $MgSO_4$, 25 mL of 1 M potassium phosphate, pH 6.0. Cool to 60°C and then add 1 mL of 1-*M* $CaCl_2$, 1 mL of 100-mg/µL carbenicillin, and 5 mL of 1 M IPTG (isopropyl-β-D-thiogalactopyranoside) (*see* **Note 2**).
11. Sytox® green (Invitrogen; Carlsbad, CA).

2.1.2. Supplies and Equipment

1. Omnitray single-well square plates (Nunc Nalgene International; Rochester, NY).
2. 96-pin replicator (Nunc Nalgene International).
3. Deep-well 96-well plates (Corning Costar; Lowell, MA).
4. Breathe-Easy sealing membranes (Sigma-Aldrich; St. Louis, MO).
5. 37°C shaking incubator with holders for 96-well plates (Fisher Scientific; Pittsburgh, PA). One holder allows stacking up to 3 deep-well 96-well plates.
6. Titertek MAP-C2 Agar Dispenser System (http://www.titertek.com).
7. Black clear-bottom half-area 96-well plates (Corning).
8. Benchtop centrifuge with 96-well plate adaptors (Beckman Coulter Allegra X-15R with Beckman Coulter SX4750 adaptors).
9. Thermo Multidrop Combi and CyBio CyBi Well Vario liquid dispensers or equivalent.
10. Vertical flow biological hood.
11 COPAS BIOSORT (Harvard Bioscience; Holliston, MA).
12 12–25°C ThermoForma worm incubators.
13 Molecular Devices Discovery-1, or equivalent automated microscope, with bright-field transmitted light option and appropriate fluorescent filter sets.
14 CellProfiler program; freely available at www.cellprofiler.org.

2.2. Chemical Screens

2.2.1. Reagents

1. LB-broth: *see* **Subheading 2.1**.
2. NGM agar: *see* **Subheading 2.1**.
3. SK-NS agar: 3 g NaCl, 3.5 g peptone, 20 g agar, 975 mL water. Autoclave. Cool to 55°C. Add in order the following sterile solutions: 1 mL of 5-mg/mL cholesterol dissolved in ethanol, 1 mL of 1 M $CaCl_2$, 1 mL of 1 M $MgSO_4$, and 25 mL 1 M potassium phosphate, pH 6.0, 1 mL of 100-mg/mL streptomycin sulfate, 1 mL 62.5 U/μL nystatin (*see* **Note 3**).
4. M9: 6 g Na_2HPO_4, 3 g KH_2PO_4, 5 g NaCl, 0.25 g $MgSO_4$–$7H_2O$, water to 1 L. Autoclave.
5. 5 M NaOH.
6. Bleach; commercially available brands are suitable.
7. *C. elegans strain glp-4(bn2);sek-1(km4) (17)*. Propagate at 15°C (*see* **Note 4**).
8. Brain-Heart Infusion (BHI) broth and agar media (BD; Sparks, MD): 52 and 37 g/L, respectively. Autoclave, cool and supplement with 100 μg/mL kanamycin sulfate.

9. *E. faecalis* strain MMH594 (cytolysin$^+$, gentamycin^R) *(18)*.
10. Bleach solution consisting of 0.1 mL of 5 M NaOH and 0.4 mL bleach.
11. 1.75X media: 35% BHI, 63.4% M9, 1.25% DMSO, 109 U/mL nystatin, and 175 μg/mL kanamycin sulfate.

2.2.2. Supplies and Equipment

1. Black-walled clear-bottom 384-well plates (Corning).
2. Gas-permeable membranes (Diversified Biotech; Boston, MA).
3. COPAS BIOSORT (Harvard Bioscience).
4. Sytox Orange (Invitrogen).
5. Bio-Tek ELx405 (Bio-Tek; Winooski, VT) or equivalent microtiter plate washer with an adjustable-height aspiration manifold.
6. Thermo Multidrop Combi (Thermo; Milford, MA) or equivalent liquid dispenser for 384-well microtiter plates.
7. Large orifice 1–200 μL pipet tips (Fisher Scientific).
8. Discovery-1 automated microscope (Molecular Devices; Sunnyvale, CA) or equivalent.
9. CellProfiler program; freely available at www.cellprofiler.org.

3. Method

3.1. RNAi Screens

3.1.1. Timeline and Throughput

The maximum number of plates manageable per experiment depends on personnel, equipment, worm strain, and particularities of the phenotype of interest. In our setting, image capture is the limiting step (Discovery-1 automated microscope; Molecular Devices). We can process up to thirty 96-well plates in 10 h (**Fig. 1**).

3.1.2. Amplify Worm Stocks

Incubation time and temperature, as well as brood size, vary with the strain. In the case of NL2099 worms, each 90-mm NGM-agar plate containing 5,000 gravid adults in their peak of egg-laying will produce enough eggs to seed 20 live hatchlings per well, in five 96-well plates (*see* **Note 1**).

3.1.3. Replicate RNAi Clones

Replicate the RNAi bacterial glycerol stocks onto agar LB-Amp/Tet single-well square plates using the following procedures.

1. Remove the tips and holders from three 200-μL tip boxes in order to use the boxes as containers.
2. Label and fill box 1 with 10% bleach, box 2 with sterile water, and box 3 with 96% ethanol.

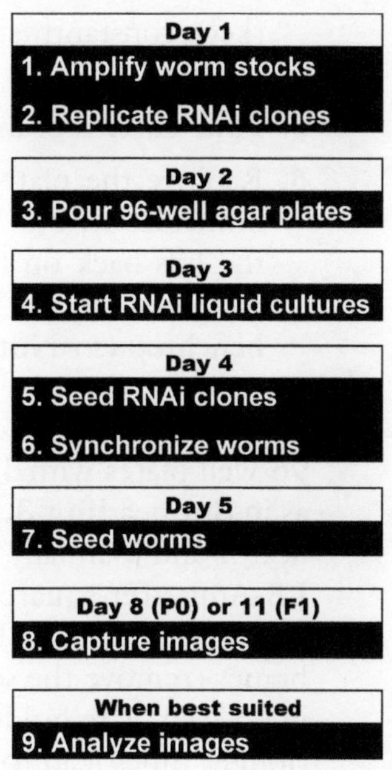

Fig. 1. RNAi screen timeline. Numeration used as in **Subheading 3.1**, where procedures are described in detail.

3. Soak the 96-pin replicator in bleach for 15 s, rinse in sterile water bath, dip in ethanol, and flame thoroughly.
4. Let the replicator cool down, firmly press onto the glycerol stocks (do not let them thaw), and then softly touch the agar plates.
5. Incubate overnight at 37°C.
6. Store agar plates with grown RNAi-feeding bacteria at 4°C (*see* **Note 5**).

3.1.4. Pour 96-Well Agar Plates

1. Prepare 20 mL of LFR media per 96-well plate, and an additional 100 mL to prime the agar dispenser. Keep agar at 60°C.
2. Autoclave 2 L of ddH_2O and keep warm at 85°C. Sterilize two 500-mL beakers with stir-bars in them.
3. Under sterile conditions, remove the lids and stack up to twenty-nine half-area 96-well plates (add a spare plate to the bottom and cover the top plate with a lid to avoid contamination). Maintain lids in sterile conditions. Use the restack function of the MAP-C2 agar dispenser to restack the plates; this function will minimize misalignment of the plates.
4. Sterilize the MAP-C2 agar dispenser tubing system by running 10 cycles with 96% ethanol. Wash out the ethanol and warm the system by running 5 cycles of hot ddH_2O. Transfer 250 mL at a time of LFR media to a sterile 500-mL beaker

(keep constantly stirring at 60°C). Prime the system with LFR media.

5. Pour 100 μL of media per well of a half-area 96-well plate.
6. Remove the plates from the stacker. This step can be done immediately, but be careful to avoid spilling the media. Put the lids back on the plates under sterile conditions. Let the media solidify for 1 h. Invert plates and leave overnight on the benchtop (*see* **Note 6**).

3.1.5. Start RNAi Cultures

Use a Thermo Multidrop Combi liquid handler to fill deep-well 96-well plates with 1.2 mL of LB-Cb. Use the same procedure as in **Subheading 3.1.3** to sterilize thoroughly the 96-pin replicator and manually replicate RNAi-feeding bacteria from agar LB-Amp/Tet square plates into deep-well 96-well plates containing 1.2 mL of LB-Cb. Seal plates with Breathe-Easy membranes (remove the second layer of the membrane). Stack plates in 96-well plate holders. Incubate for 8–14 h at 37°C, 200 rpm (longer times lead to reduced efficiency) (*see* **Note 7**).

3.1.6. Seed RNAi Clones

1. Label or barcode all plates.
2. Centrifuge the cultures grown in deep-well 96-well plates for 5 min at 2,000 × g.
3. Remove the supernatant (*see* **Note 8**).
4. Using the Thermo Multidrop Combi liquid handler with a standard cassette set to dispense into a deep-well 96-well plate, add 100 μL of S-buffer/well. Using a CyBio CyBi Well Vario with a 96-well head adaptor, resuspend bacteria using 5 cycles of pipetting at speed of 300 rpm. Seed 25 μL of bacteria in two independent plates (duplicates). Leave plates on benchtop until drying starts (*see* **Note 9**).
5. Dry plates in vertical-flow biological hood for 14 h (*see* **Note 10**).

3.1.7. Synchronize Worms

1. Harvest gravid adults and laid eggs of up to four 90-mm plates with 2.5 mL of S-buffer per plate, and transfer them to 15-mL conical tubes. Centrifuge at 1,500 × g for 30 s to pellet worms and eggs (all centrifugation steps are performed using the same conditions). Remove the supernatant by aspirating the liquid with a sterile glass Pasteur-pipette attached to a vacuum line (every time the supernatant is removed, it should be done this way). Resuspend the pellet in 3 mL of S-buffer.
2. Add 3 mL of freshly prepared lysis solution. Shake and vortex the tube for 1 min. Quickly fill tube with S-buffer and centrifuge. Remove 12 mL of supernatant. Add 3 mL of lysis solution to the 3 mL of lysate remaining, vortex and shake for 1–2 min. Quickly fill tube with S-buffer and centrifuge. Remove the

supernatant, leaving as little volume as possible without disturbing the pellet. Repeat the washing step (fill the tube, centrifuge, and remove all the supernatant) at least three times. Do all washing steps under sterile conditions (*see* **Note 11**).

2. Resuspend the eggs and embryos in 5 mL of S-buffer and let them hatch for no less than 18 h at RT with gentle rocking. Hatched worms will arrest at the first larval stage (L1) in the absence of food.

3.1.8. Seed Worms

Worm Preparation in order to separate worm and agar debris from L1 larva, centrifuge for 3 min at 150 × g. Transfer the supernatant containing the synchronized L1s to a fresh tube (leave the bottom 250 μL of liquid and discard the tubes). Pool suspensions and dilute or concentrate to have 10 L1s/μL. At least 1 h before seeding the L1s, add 1 μM Sytox® green (final concentration) (*see* **Note 12**). Leave worms gently rocking at RT until seeding. The combination of Sytox® green with the use of a worm sorter (COPAS Biosort) allows dispensing only live L1s in the 96-well plates.

1. COPAS Biosort Preparation: Sterilize the worm sorter tubing system by running 200 mL of 10% bleach. Wash out the bleach by running 200 mL of sterile ddH_2O. Run 200 mL of 70% ethanol. Wash out the ethanol by running 200 mL of sterile ddH_2O. Equilibrate with sample buffer, by running 50 mL of S-buffer.
2. Add 200 mL of S-buffer and 2 mL of Sytox® green-stained L1 suspension to the sample cup. Perform an acquisition cycle of 500 objects. For accurate dispensing of L1s, the flow rate should be approximately 10 events/sec. Dilute or add L1 suspension accordingly. Based on size and intensity of the fluorescent signal, define the gate of interest (**Fig. 2**). Perform a test by dispensing 5 non-green objects (live L1s) per well in the cover of a 96-well plate. Check accuracy under the dissecting scope (*see* **Note 13**).

3.1.8.1. Sorting for one-Generation Screens

Seed 20 live (non-green) L1s per well of a 96-well plate. Dry plates for 1 h in biological hood.

3.1.8.2. Sorting for Two-Generation Screens

Seed two live (non-green) L1s per well of a 96-well plate.

3.1.9. Incubation

1. One-generation screens: Incubate for 3 days at 20°C, then shift to 25°C. Next day, worms on control RNAi will be adults laying unfertilized oocytes (thermosterile background).
2. Two-generation screens: Incubate for 84 h at 20°C and then transfer to 25°C. At this time NL2099 worms on control RNAi will have laid approximately 10 eggs. After shifting to 25°C, NL2099 worms will lay approximately 20 more eggs before they start laying unfertilized oocytes. Maintain the

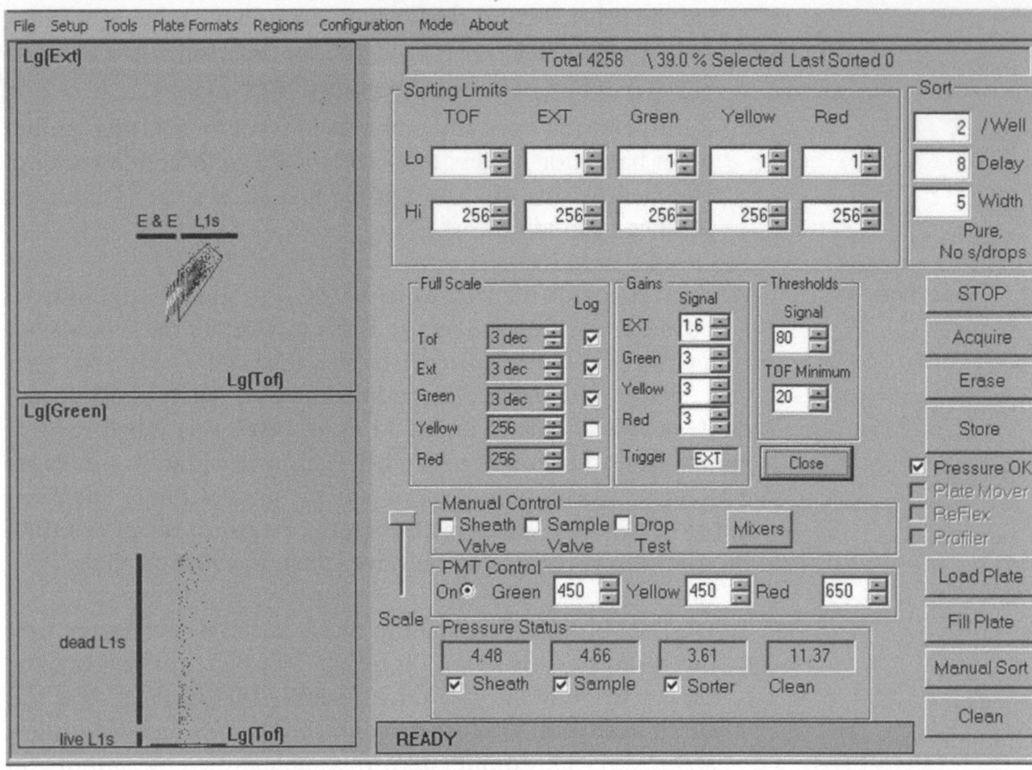

Fig. 2. Optimized COPAS Biosort conditions for sorting L1s. The post-run interface is depicted. The *upper dot-plot* shows the distribution of a post-egg prep population. The extinction or density (EXT) vs. the time of flight or length (TOF) is graphed. The *gated area* includes mostly L1s (dead and live) and excludes most of the non-hatched eggs and embryos (E & E). The *lower dot-plot* shows the distribution of the L1 subpopulation when the TOF vs. the green fluorescent intensity is graphed. The gated area includes live (*non-green*) L1s. In addition a tail of dead (Sytox® green positive) L1s is observed. Using the depicted settings, two live L1s, *non-green* gated objects, are sorted per well.

cultures at 25°C until the day on which scoring will occur. For most RNAi clones, the second generation will reach adulthood without starving (*see* **Note 14**).

3.1.10. Image Capture

Microscope: The major challenge to automate image capture in this format is imposed by the use of agar media. The background has high intensity and the focal plane changes from well to well because the media dries at different rates. An instrument able to perform image-based focusing is therefore indispensable. Capture settings will be different for different instruments and they will need to be adjusted for different 96-well plates and samples. The optimized parameters to capture bright-field, green-, and red-fluorescent signal with a Discovery-1 automated microscope (Molecular Devices) are described in **Table 1**. This setup allows imaging a half-area 96-well plate at three different wavelengths (bright-field, green-, and red-fluorescent signal) in 15 min (*see* **Note 15**).

Table 1
Image capture settings

Parameter	Setting
Plate Reference Point (a)	14337.5
Reference Objective (b)	5
Parfocality Offset (c)Offset	1,445
Plate Height (d)	14.2
Well Depth (e)	10,500
Find 2nd Maximum (f)	FALSE
Start z position (g)	19,982.5
Full range (μm) (h)	1,000
Full max step (μm) (i)	590
Plate bottom exposure (j)	3
Wide range (μm) (k)	50
Wide max step (μm) (l)	10
Accuracy (μm) (m)	59
Magnification	2×
Camera binning	2
Gain	2 (4×)
Transmitted light exposure	10 ms
Image based range (μm)	500
Max. step (μm)	100
Nile Red: filter set	572/630
exposure time	100 ms
GFP: filter set	470/530
exposure time	250 ms

Settings used to capture images of worms growing on half-area 96-well agar media plates with a Discovery-1 automated microscope are listed. Images are taken in 3 wavelengths: bright-field, GFP, and Nile Red.
(a) Reference point of flat sheet in plate holder. Value is distance from application z origin.
(b) Objective position used for setting reference point.
(c) Offset distance between current objective to reference-point objective.
(d) Height defined for current plate.
(e) Depth of well for current plate.
(f) TRUE = two-peak search, FALSE = single-peak search3
(g) Start z position of search in units.
(h) Total range covered in μm.
(I) Incremental steps in μm.
(j) Image exposure (ms).
(k) Search range at bottom of well in μm.
(l) Incremental steps in μm.
(m) Accuracy to which focus will be found (μm).

Samples on the scoring day, transfer plates to be documented to 12°C. This step will slow down growth, allowing the scoring of animals of the same age, and will help prevent starvation in two-generation screens. Take plates out of the incubator one at a time and capture images (*see* **Note 16**).

3.1.11. Image Analysis

Analysis is performed with CellProfiler (*see* **Chap. 15**). An optimized pipeline for green, red, and bright-field worm area analysis is available at www.cellprofiler.org.

3.1.12. Replication

RNA interference by feeding is highly variable; therefore, it is imperative to retest any putative phenotypes observed. We recommend retesting in 96-well plates and larger plate formats, so that a larger number of worms can be scored and phenotypes confirmed.

3.2. Chemical Screens

The protocol described below allows the screening of twenty 384-well plates per experiment. The rate-limiting steps are worm dispensing and imaging. Each of the limiting steps takes approximately 15 min per plate (**Fig. 3**).

3.2.1. Amplify Worm Stocks

1. Inoculate one colony of *E. coli* HB101 into a 2-L flask containing 500 mL LB supplemented with 200 μg/mL streptomycin sulfate. Incubate for 16 h at 37°C and 250 rpm. Centrifuge the saturated culture at 5,000 × g for 10 min and resuspend in

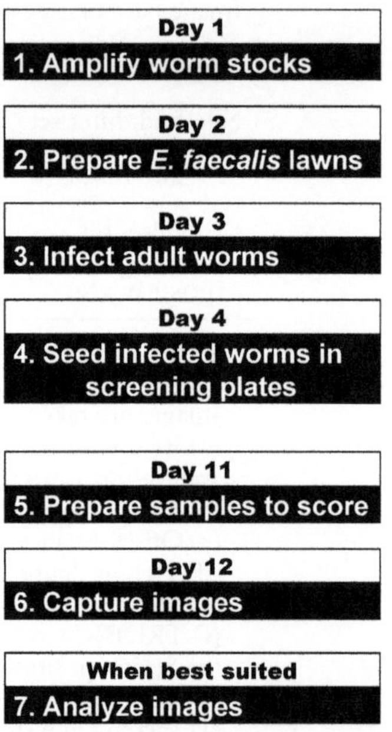

Fig. 3. Compound screen timeline. Numeration used as in **Subheading 3.2**, where procedures are described in detail.

LB to concentrate the bacteria 20-fold. Spread 100 μL of the concentrated bacteria onto 90-mm plates of NGM or SK-NS agar. Incubate the plates at RT for 1 day to produce the lawns of *E. coli* that will serve as the food source for the worms. Plates can be stored at RT for up to 1 week.

2. Grow approximately 1,000 *glp-4(bn2ts);sek-1(km4)* L1 worms on each 90-mm plate of HB101 on NGM at 15°C for 5 days until the worms become gravid adults.
3. Harvest the gravid adults by washing the plates with 10 mL of M9 and transfer them into 15-mL conical tubes. Centrifuge the tubes at 1,500 × g for 30 s to pellet the worms. Remove the supernatant, leaving behind 0.5 mL of liquid and the sedimented worms. Immediately before use, add 0.5 mL bleach solution to the tube. Continuously agitate and vortex the tube for 3–5 min, until most of the adult worms have ruptured. Do not overexpose to the bleach solution since this will result in damaged embryos. An average of ten to fifteen progeny per *glp-4(bn2ts);sek-1(km4)* adult will be obtained.
4. Wash the eggs 3 times with 14 mL M9 (*see* ***Subheading 3.1.7, steps 1–2** for wash procedures).
5. Resuspend the eggs in 5 mL M9 and let them hatch at RT for 20 h with gentle rocking. The hatched worms arrest at the L1 stage in the M9 buffer. Seed approximately 4,000 L1s onto each HB101 on SK-NS plate and grow the worms at 25°C for approximately 54 h until the worms become young adults.

3.2.2. Prepare E. faecalis Lawns

To prepare lawns of the pathogenic *E. faecalis* bacteria as a source of infection, inoculate one colony of *E. faecalis* strain MMH594 in BHI liquid media and incubate at 37°C for 6 h. Spread 100 μL of the culture over the entire surface of 90-mm plates containing BHI agar. Incubate the plates at 25°C overnight to grow the lawns and then cool the plates at 15°C.

3.2.3. Infect Adult Worms

To each plate of sterile, adult worms, add 15 mL of M9 and resuspend the worms by gently shaking the plate for 10 s. Transfer the worms into sterile 50-mL tubes. Allow the worms to settle to the bottom of the tubes and remove the supernatant. Wash worms twice with M9 to remove the *E. coli.*

1. Using large-orifice pipette tips, seed the worms onto the lawns of *E. faecalis* MMH594 on BHI agar. Transfer up to 8,000 animals per plate. Incubate the plates for 15 h at 15°C to allow the infection to become established (*see* **Note 17**).

3.2.4. Seed Infected Worms in Screening Plates

1. Using a multiplate dispenser, dispense 20 μL of 1.75X media to each well of the 384-well plate.

2. Pin transfer 100 nL of a 5-mg/mL compound stock solution (dissolved in DMSO) into each well of a 384-well plate (final DMSO concentration is 1% and compound concentration is 14 μg/mL).
3. Sterilize and prepare COPAS Biosort as described in **Subheading "COPAS Biosort preparation"** 3.1.8.1***.
4. Resuspend the infected worms in M9. Dispense 15 young adult worms into each well of the 384-well plates using the COPAS Biosort (*see* **Note 18**).
5. Dry the top of the 384-well plates with laboratory tissue to allow adhesion of membranes and seal the plates with gas-permeable membranes.
6. Incubate plates at 26.5°C with 85% relative humidity for 7 days (*see* **Note 19**). Place the plates in a single layer on top of the shelves and incubate without agitation.

3.2.5. Prepare Samples to Score

1. Resuspend the worms and bacteria by vortexing the plates for 5 s at a high setting. Centrifuge the plates at 1,000 × g for 10 s to remove the liquid from the membranes. Remove the membrane seals. Using a plate washer, dispense 65 μL of M9 per well using the maximum dispense speed to facilitate washing. Let the worms settle to the bottom of the wells for at least 3 min. Remove three-fourths of the liquid from the top of the plate using the aspirating head of the plate washer. Wash the plates a total of four times. After the final wash, aspirate enough liquid to leave ~ 25 μL/well (*see* **Note 20**).
2. Using the multiplate dispenser, dispense 50 μL of 1-μM Sytox Orange (diluted in M9) per well resulting in a stain concentration of 0.7 μM.
3. Seal the plates with gas-permeable membranes and incubate at 20°C for 20 h at 85% relative humidity. Arrange the plates in a single layer on the incubator shelves and incubate without agitation.

3.2.6. Image Capture

Wells are imaged using a Molecular Devices Discovery-1 automated microscope. Fluorescent TRITC (535 nm excitation, 610 nm emission) and transmitted light images are captured. Hardware capture conditions are as described in **Table 1**, except that the focal plane is set at the bottom of the well (laser based scanning). The use of a 2× objective allows capturing the entire area of a well in a 384-well plate.

3.2.7. Image Analysis

Analysis is performed with CellProfiler (*see* **Chap. 15**). An optimized pipeline to quantify worm survival (1 – (Sytox Orange worm area/bright-field worm area) = 1 – (dead worm area/total worm area)) is available at www.cellprofiler.org. An example of the quantification of worm survival using the described methodology is shown in **Fig. 4** (*see* **Note 21**).

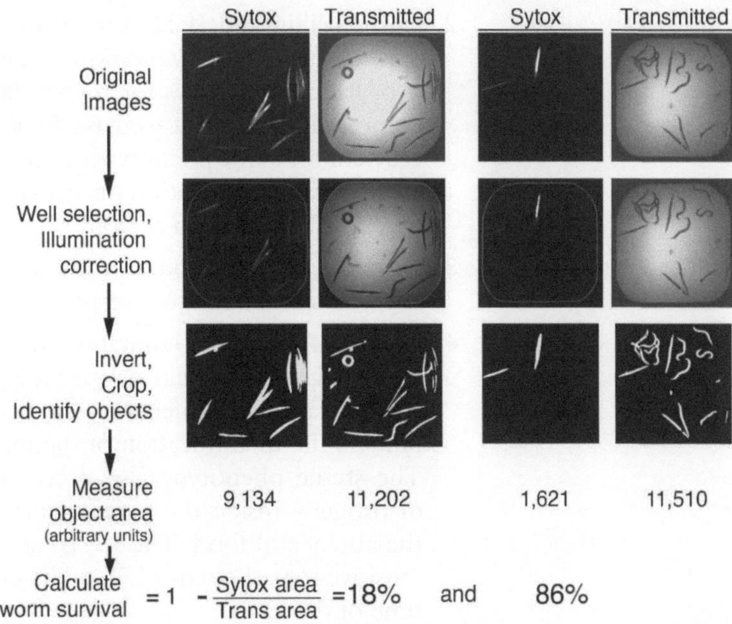

Fig. 4. Worm survival quantification using CellProfiler. Worms in 384-well plates are incubated with Sytox® orange, which specifically stains dead worms. *Top row* shows raw fluorescent Sytox® orange and bright-field images (captured using a ×2 objective) of an untreated well and an antibiotic-treated well. The images were analyzed using CellProfiler through a pipeline of 29 processing steps. The results of two of the processing steps are shown. Images in the *middle row* show the results of well-area selection and light correction of the bright-field images. Images in the *bottom row* show the result of the worm-identification function after inverting and cropping the well area. Finally, the total object area of the fluorescent and the bright-field images are measured. The areas of the Sytox® orange objects and bright-field objects are used to approximate the number of dead worms and the total number of worms, respectively. Worm survival = 1 – (Sytox® orange area/total bright-field worm area).

4. Notes

1. Finding the optimal worm genetic background is one of the most important aspects of developing a robust assay that can then be automated. We chose NL2099 because it carries a mutation in an RNA-dependent RNA polymerase (RRF-3), which makes this strain hypersensitive to RNAi *(16)*. The RRF-3 deficiency promotes stronger and additional loss-of-function phenotypes. In addition, *rrf-3* makes animals sterile at temperatures ≥ 25°C. The thermosterility allows the control of the number of progeny by shifting animals to 25°C. NL2099 does not have a significantly smaller brood size at 20°C when compared to wild-type animals, yet (as explained in **Subheading 3.1.8**) accurate dispensing of worms requires a large excess of synchronized progeny.

2. Carbenicillin and IPTG should not be stored for more than 2 weeks at 4°C. If starvation becomes a problem, adding 2 g of lactose/L allows for thicker lawns (S. Fischer, personal communication), but we have found that RNAi becomes less efficient for some phenotypes under these conditions. If interested in fat accumulation phenotypes, add 1/200,000 of 0.5-mg/mL Nile Red (in acetone).
3. Nystatin and streptomycin are added to inhibit fungal and bacterial contamination, respectively.
4. The *glp-4(bn2)^{ts}* *(19)* mutation makes the worms sterile when they are grown at the restrictive temperature of 25°C. Sterility is important in the infection assay because progeny production hinders the quantification of the initial population of worms. The sterile phenotype also prevents the matricidal hatching of progeny inside the parent worm that normally occurs in the absence of food. The *sek-1(km4)* mutant worms are more sensitive to pathogens *(17)* and its use shortens the incubation time of the assay.
5. Cross-contaminating RNAi-feeding bacteria is one of the most serious problems to which mishandling of stocks can lead. Extreme caution should be taken to thoroughly sterilize the replicator between plates. To maximize RNAi efficiency: Do not store replicated clones for more than four weeks; Do not replicate from agar plate to agar plate, or from liquid cultures to agar plates.
6. Alternatively, agar media can be manually dispensed. Keep media stirring at 60°C. Remove from two benchtop thermoblocks the metal racks normally used to hold tubes, thoroughly clean them with ethanol, and set them at 85°C. Place the 96-well plate to be filled with media in one thermoblock and the reservoir containing the media in the second block. Using a 12-channel pipettor, dispense 100 μL of media per well. Preheat the tips by slowly pipetting up and down 100 μL of media (two times). Dispense 100 μL of media slowly and carefully. Do not eject the last remaining bit of media to minimize bubble formation. Change tips with every plate or when bubbles appear.
7. Alternatively, use a 12-channel electronic pipettor and 1.2-mL sterile tips to fill deep-well 96-well plates with 1.2 mL of LB-Cb. Do not use less than 1.2 mL because the pin replicator will not touch the liquid if less volume is used. Alternatively, dispense 1 mL of LB-Cb and inoculate by using a multichannel pipettor.
8. In order to minimize contamination and/or cross-contamination of the samples, thoroughly clean, with bleach and ethanol, the sink area, and place three layers of clean paper towels next

to the sink. After centrifugation, quickly invert the plates to dispose the supernatant in the sink containing bleach, place the plates inverted on the paper towels to absorb the excess of liquid, and leave for 5 min.

9. Alternatively, using a 12-channel pipettor, manually add 100 μL of S-buffer to all wells of the deep-well plates containing the bacterial pellets. Resuspend bacteria and add 25 μL of RNAi bacteria per well in two agar 96-well plates (duplicate RNAi clones).

10. Drying conditions may vary from hood to hood. We found that adding 10 μL extra of water to columns 1 and 12, and to rows A and H, helps prevent the cracking of the media. We do not recommend drying in a horizontal air-flow hood because drying will be very uneven, leading to cracked outer wells and wet inner wells.

11. Different labs use different synchronization protocols. We present two versions in this chapter. **Subheading 3.1.7** describes a double-bleaching protocol that helps to dissolve most of the carcasses that could later interfere with dispensing or imaging of the worms. In addition, **Subheading 3.1.7** uses S-buffer instead of M9. The S-buffer salt content is much lower than that of M9. If M9 is used in half-area 96-well plate assays, the final concentration of salts after drying the plates is too high and appears to affect worm growth rates.

12. Sytox® green is a membrane-impermeable dye that stains nucleic acids in dead and dying cells. In an L1 population synchronized by bleaching, it will stain dead eggs, embryos, and L1s.

13. In our hands, 5% of the wells will have ±1 worm. When dispensing fewer than five L1s per well, use the "pure" mode. The use of the pure mode allows maximum accuracy, but it requires a large excess of animals because most of the objects will not fulfill the sorting specifications (*see (20)* for more details on sorting conditions and requirements). When dispensing five or more L1s per well use the "enhanced" mode.

14. A caveat of this approach is that the second generation of the slow-growing RNAi clones will not be scorable because plates will be shifted to 25°C before they would have started laying eggs.

15. Useful general tips: Use a different starting z-position for external (rows A and H or columns 1 and 12) and internal wells; this will greatly reduce scanning time. Focus using bright-field imaging; this step will minimize the fleeing

behavior normally triggered in worms by short-wavelength light. Do not apply background subtraction or shadow correction; modifying raw images will have a negative impact on subsequent analysis.

16. If signal from a transgene is the readout of your assay, be aware that in some cases low temperatures can lead to transgene silencing.
17. Worms are infected on the lawns of pathogen for a period of time that allows persistent intestinal colonization, but at which symptoms are not yet observed.
18. The worm synchronization method described in **Subheading 3.2.1**, steps 3–4, results in a population of worms of which 95% are young adults. Based on TOF, the COPAS Biosort differentiates young adults from younger animals (**Fig. 2**). Fifteen young adults are dispensed per well of a 384-well plate. The remaining 5% of the worm population corresponds to slower-growing animals that are discarded. The COPAS Biosort transfers each worm in a volume of ~1 μL. The final volume per well is 35 μL. Final concentrations are as follows: 20% BHI, 36% M9, 1% DMSO, 100 μg/mL kanamycin sulfate, 62.5 U/mL nystatin, and the remaining liquid consists of sheath fluid (worm sorter specific fluid *(20)*) and M9.
19. Roughly 90% of the untreated worms die from the infection during the 7-day incubation period. In contrast, less than 15% of the worms die when treated with antibiotics such as ampicillin or tetracycline. Humidity is set at the maximum of the incubator capacity to reduce evaporation. Alternatively, the microtiter plates could be placed into containers that are lined with wet paper towels.
20. In addition to staining dead worms, Sytox® Orange also stains the bacteria in the assay wells and therefore the bacteria need to be removed to allow quantification of the fraction of dead worms.
21. We quantify survival by measuring areas (number of pixels) instead of counting worms because CellProfiler cannot distinguish overlapping worms as independent objects. The decision to measure area occupied by dead worms, instead of fluorescent Sytox Orange intensity was based on the observation that the intensity of the fluorescent-staining dead worms varied greatly. This variation, in addition to the inhomogeneous nature of the worm suspensions, also prevented the use of a fluorescent plate reader.

Acknowledgments

The authors thank Fred Ausubel, Harrison Gabel, and Jonah Larkins-Ford for critical review of the manuscript. We also thank Nicola Tolliday, Bridget Wagner, Lynn Verplank, Jason Burbank, and Anne Carpenter for intellectual and practical contributions to the development of the assays. This work was supported by a grant from the SPARC Program (Broad Institute of Harvard and MIT) to Gary Ruvkun, Fred Ausubel, Eyleen O'Rourke, David Sabatini, and Jonathan Clardy, and by NIH grant AI072508, Fred Ausubel PI.

References

1. Levitan D, Greenwald I. Facilitation of lin-12-mediated signalling by sel-12, a *Caenorhabditis elegans* S182 Alzheimer's disease gene. *Nature* 1995;377(6547):351–354.
2. Ogg S, Paradis S, Gottlieb S, et al. The Fork head transcription factor DAF-16 transduces insulin-like metabolic and longevity signals in *C. elegans*. *Nature* 1997;389(6654):994–999.
3. Ranganathan R, Sawin ER, Trent C, Horvitz HR. Mutations in the *Caenorhabditis elegans* serotonin reuptake transporter MOD-5 reveal serotonin-dependent and -independent activities of fluoxetine. *J Neurosci* 2001;21(16):5871–5884.
4. Lakso M, Vartiainen S, Moilanen AM, et al. Dopaminergic neuronal loss and motor deficits in *Caenorhabditis elegans* overexpressing human alpha-synuclein. *J Neurochem* 2003;86(1):165–172.
5. Bergamaschi D, Samuels Y, O'Neil NJ, et al. iASPP oncoprotein is a key inhibitor of p53 conserved from worm to human. *Nat Genet* 2003;33(2):162–167.
6. Ashrafi K, Chang FY, Watts JL, et al. Genome-wide RNAi analysis of *Caenorhabditis elegans* fat regulatory genes. *Nature* 2003;421(6920):268–272.
7. Cronin CJ, Feng Z, Schafer WR. Automated imaging of C. *elegans* behavior. *Methods Mol Biol* 2006;351:241–251.
8. Kamath RS, Fraser AG, Dong Y, et al. Systematic functional analysis of the *Caenorhabditis elegans* genome using RNAi. *Nature* 2003;421(6920):231–237.
9. Timmons L, Fire A. Specific interference by ingested dsRNA. *Nature* 1998;395(6705):854.
10. Lee SS, Lee RY, Fraser AG, Kamath RS, Ahringer J, Ruvkun G. A systematic RNAi screen identifies a critical role for mitochondria in C. *elegans* longevity. *Nat Genet* 2003;33(1):40–48.
11. Frand AR, Russel S, Ruvkun G. Functional genomic analysis of *C. elegans* molting. *PLoS Biol* 2005;3(10):e312.
12. Kim JK, Gabel HW, Kamath RS, et al. Functional genomic analysis of RNA interference in *C. elegans. Science* 2005;308(5725):1164–1167.
13. Lehner B, Crombie C, Tischler J, Fortunato A, Fraser AG. Systematic mapping of genetic interactions in *Caenorhabditis elegans* identifies common modifiers of diverse signaling pathways. *Nat Genet* 2006;38(8):896–903.
14. Kwok TC, Ricker N, Fraser R, et al. A small-molecule screen in *C. elegans* yields a new calcium channel antagonist. *Nature* 2006;441(7089):91–95.
15. Moy TI, Ball AR, Anklesaria Z, Casadei G, Lewis K, Ausubel FM. Identification of novel antimicrobials using a live-animal infection model. *Proc Natl Acad Sci U S A* 2006;103(27):10414–10419.
16. Simmer F, Tijsterman M, Parrish S, et al. Loss of the putative RNA-directed RNA polymerase RRF-3 makes *C. elegans* hypersensitive to RNAi. *Curr Biol* 2002;12(15):1317–1319.
17. Kim DH, Feinbaum R, Alloing G, et al. A conserved p38 MAP kinase pathway in *Caenorhabditis elegans* innate immunity. *Science* 2002;297(5581):623–626.
18. Shankar N, Coburn P, Pillar C, Haas W, Gilmore M. Enterococcal cytolysin: activities and association with other virulence traits in a pathogenicity island. *Int J Med Microbiol* 2004;293(7–8):609–618.
19. Beanan MJ, Strome S. Characterization of a germ-line proliferation mutation in *C. elegans. Development* 1992;116(3):755–766.
20. Pulak R. Techniques for analysis, sorting, and dispensing of *C. elegans* on the COPAS flow-sorting system. *Methods Mol Biol* 2006;351:275–286.

Chapter 6

Whole-Organism Screening: Plants

April Agee and David Carter

Summary

The small plant *Arabidopsis thaliana* has been an indispensable tool for plant biologists working in fields that utilize cell biology, molecular biology, and genetics; these topics are almost universal in plant biology studies, ranging from genomics to ecology. In this chapter, we present a start-to-finish approach to high-throughput imaging of Arabidopsis that caters to two different audiences: those who are working with plants for the first time, and plant scientists looking to the apply high-throughput imaging to existing projects.

Key words: Arabidposis, High-content screen, Plant screening, Whole organism.

1. Introduction

The small plant *Arabidopsis thaliana* (thale cress or mouse-ear cress) has been an indispensable tool for plant biologists working in fields that utilize cell biology, molecular biology, and genetics; these topics are almost universal in plant biology studies, ranging from genomics to ecology. The beneficial features of this organism are well-known in the plant community. Arabidopsis (now the common name) shares many of the qualities of other model organisms, such as a small genome, short generation time, prolific production of offspring, easy transformation, and, most importantly, similarity to other species with more direct applications, such as rice, soybean, maize, and cotton. Further, multinational research efforts have led to the full sequencing of the Arabidopsis genome, the production of extensive genetic and physical maps of all five chromosomes, and the generation and characterization of

P.A. Clemons et al. (eds.), *Cell-Based Assays for High-Throughput Screening, Methods in Molecular Biology, Vol. 486*

DOI: 10.1007/978-1-60327-545-3_1

Table 1
Resources and suppliers of Arabidopsis materials

The Arabidopsis Information Resource (TAIR, http://www.arabidopsis.org)	Database of genetic and molecular biology data, including seed stocks from sources mentioned below
TAIR Map Viewer (http://www.arabidopsis.org/servlets/mapper)	Arabidopsis genome map viewer
The Institute for Genomic Research (TIGR, http://www.tigr.org/plantProjects.shtml)	Genomic sequence data and annotation generated at TIGR and assemblies of Arabidopsis ESTs
The Arabidopsis Biological Resource Center (ABRC, http://www.biosci.ohio-state.edu/plantbio/Facilities/abrc/abrchome.htm)	Arabidopsis seed stock center at Ohio State University
The Nottingham Arabidopsis Stock Centre (NASC, http://arabidopsis.info)	European Arabidopsis seed stock center
Salk Institute Genomic Analysis Laboratory (SIGnAL, http://signal.salk.edu/)	Sequence-Indexed Library of Insertion Mutations in the Arabidopsis Genome
The Arabidopsis Book (TAB, http://www.aspb.org/publications/arabidopsis/)	The American Society of Plant Biologists' online information resource reviewing important topics in Arabidopsis research in detail
German Plant Genomics Program of BMBF Kölner Arabidopsis T-DNA lines (GABI-KAT, http://www.gabi-kat.de/)	Collection of T-DNA mutagenized Arabidopsis lines
pEarleyGate Vectors (http://www.biology.wustl.edu/pikaard/pEarleyGate%20plasmid%20vectors/pEarleyGate%20homepage.html)	Gateway (Invitrogen)-compatible vectors for engineering and expressing epitope-tagged or fluorescent recombinant proteins in plants

a large number of mutant lines, which are readily available from public stock centers (**Table 1**). Those new to Arabidopsis research will benefit from Detlef Weigel & Jane Glazebrook's comprehensive laboratory manual for Arabidopsis, including sections on plant growth, genetic analysis, proteomics, and genomics *(1)*.

Visualization of protein localization within individual cells as well as protein expression within plant tissues can provide valuable information about molecular function. Proteins that colocalize to the same set of subcellular structures often have roles in similar biological processes, and localization in highly specialized tissues like pollen or guard cells may provide additional clues to functionality. Consortium efforts, including the National Science Foundation's Arabidopsis 2010 project, which aims to determine the functions of all Arabidopsis proteins, have approached this daunting task by developing fluorescent fusion proteins for

Table 2
Typical timings for preparing and scanning Arabidopsis on plate lids

Procedure	Time
Dispensing agar	3–4 min
Dispensing seeds	15 min
Incubation	7 days
Scanning, Epson	3 min
Scanning, Typhoon @ 50um	13 min
Hit Picking, 48 wells	~15 min
Imaging DC-48 @ 5 hits/well	32 min
Imaging lids @72–150/plate	200 hit sites per hour
Quantifying seedling shape	~10 min for 48

To minimize stress on the seedlings, all imaging is completed within 1 h of placing the cover glass. At a Typhoon resolution of 50 μm, five "hit" sites per seedling can be screened for all 48 seedlings. In mutant screens, seedlings can be grown at a density of 150 per plate, and mutants of interest can be rescued to agar, then soil

known and unknown Arabidopsis proteins. These marker lines are now being used to answer more complex biological questions about compartments within plant cells, and are freely available at the Arabidopsis Biological Resource Center (ABRC; **Table 1**).

The power of these new marker lines comes with the identification of genes or chemicals affecting marker localization; however, traditional microscopy is not suited for the immense number of images required for such studies. Therefore, this system for large-scale high-throughput confocal imaging of plants was developed (**Table 2**). It has already proven itself to be invaluable for determining localization of known marker proteins in response to mutations affecting vacuole structure, cell shape, endomembrane protein transport *(2)*, and small-molecule response *(3)*. As new marker lines are developed, a fast system for fluorescence screening has become even more essential for answering questions in both basic and applied plant biology.

For a detailed review of GFP technology and a discussion of other reporter genes, we suggest Brandizzi and colleagues' review *(4)*. In short, reporter systems such as β-glucuronidase (GUS) and luciferase have been used with success in plant gene-expression studies, but both have serious limitations in high-throughput imaging for genetics and chemical-genomics applications. Although extremely precise, histochemical analysis of GUS expression is lethal to the plant, which limits its applications for genetics. Luciferin-luciferase imaging allows for in vivo imaging of low-abundance gene expression in plants, but high-performance luminescence imagers are not widely available. Poor resolution and low signal strength have limited adoption of luminescence for large-scale and high-throughput imaging. In contrast, fluorescent markers such as the green fluorescent protein from the jellyfish *Aequoria victoria* *(5)* and its spectral derivatives required some initial tweaks for optimal expression in plants *(6)*, but have emerged as the dominant technology in fusion-protein studies. This emergence is due to the rapid and accurate in vivo localization of chimeric proteins by confocal fluorescence microscopy.

Choosing a stable, robust marker line whose protein localization has been confirmed is critical for any large-scale study of an organelle or tissue. Any chimeric protein has the potential to be mistargeted and thus to inaccurately represent the true localization of a protein of interest. Protein tags can obscure a targeting signal on the N- or C-terminus of the peptide *(7)*, or lead to misfolding or novel intracellular interactions. In addition, strong constitutive promoters are often used to enhance visualization of proteins with low levels of expression, but this technique can lead to expression in novel tissues or organelles or, potentially worse, an overexpression phenotype. We mention this caveat to emphasize the importance of marker choice for high-throughput imaging studies.

Sean Cutler and colleagues' study *(8)* used random GFP:cDNA fusions under the control of the strong 35S promoter from cauliflower mosaic virus to examine subcellular protein localization in plants (**Fig. 1**). Over 100 lines expressed GFP in distinct compartments that were discernable from soluble GFP. These lines are publicly available from ABRC (**Table 1**), and at least one line has proven successful for high-throughput imaging of mutations which alter endomembrane-protein transport or vacuole structure *(2)*. Although numerous robust organelle markers have originated from these cDNA fusions, later studies examined localization of native genes controlled by their own promoters using a technique known as Fluorescent Tagging of Full Length Proteins, or "FTFLP" *(7, 9)*. FTFLP uses an internal fluorescent tag to minimize alteration of protein targeting by N- and C- terminal fusions. Forty distinct subcellular compartments have been identified in Arabidopsis with this technique, and projects are ongoing to predict localization based on existing data for known proteins and computational analysis *(9)*. These marker lines are also available from ABRC (**Table 1**).

Although the FTFLP *(7)* and random cDNA fusion *(8)* techniques have generated markers for almost all known plant cell compartments, researchers often want to study localization of a specific protein fusion. Either of these techniques can be duplicated; we have also found success with Keith Earley and colleagues' *(10)* set of Gateway-compatible "pEarleyGate" vectors

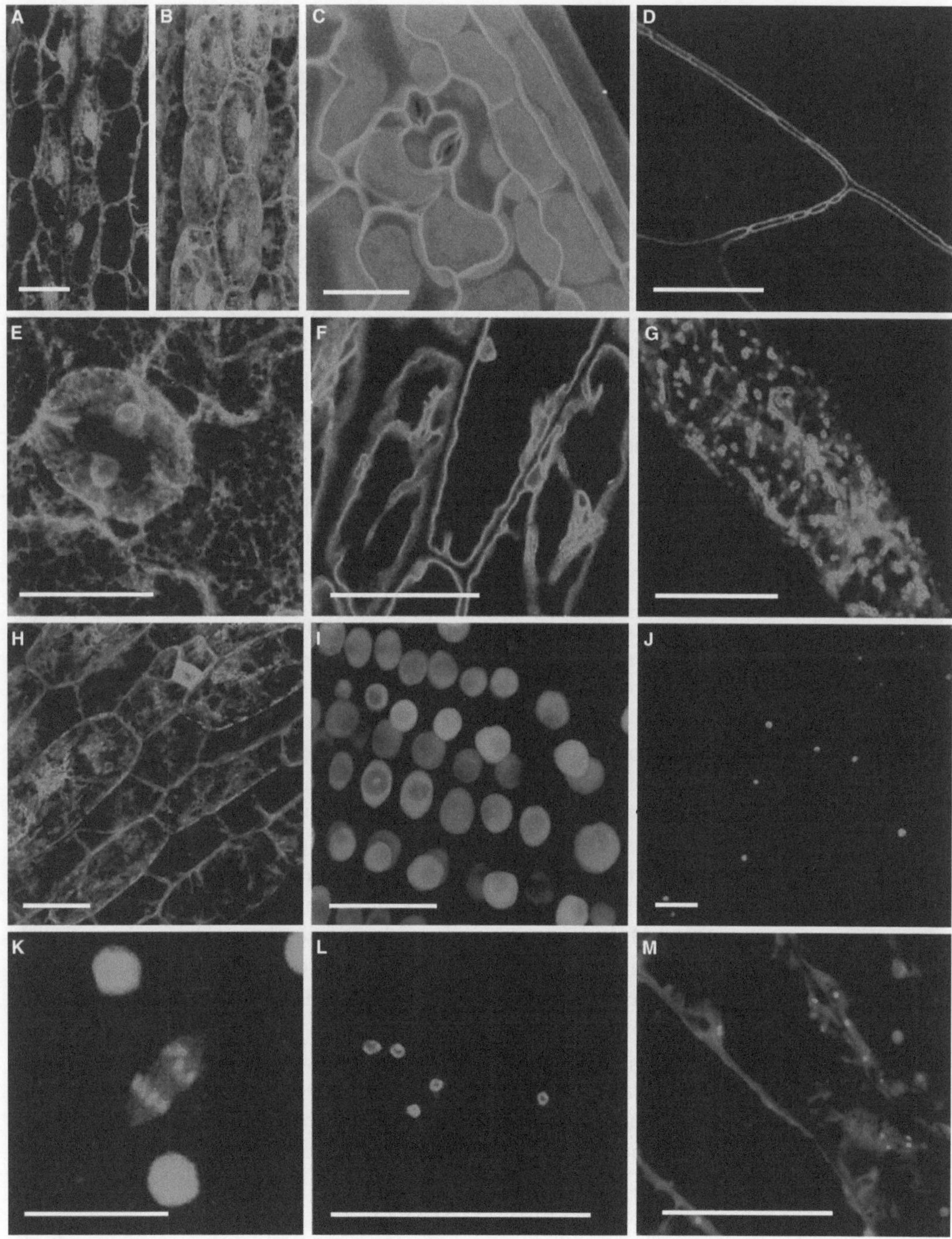

Fig. 1. GFP marker lines are available which indicate a wide range of sub-cellular localizations: (A, B) cytosol, with nuclei; (C, H) cytosol, excluded from nuclei; (D) plasma membrane; (E) endoplasmic reticulum; (F) vacuolar membrane; (G) unknown cytosolic vesicles; (I) nuclei; (J) nucleoli; (K) chromosomes; (L, M) dots of unknown identity. Bars = 20 μm. Adapted with permission from *(8)*.

(Invitrogen) for functional genomics in plants. These vectors are available through ABRC, and detailed instructions for use are provided at the pEarleyGate webpage (*see* **Note 1**). The pEarleyGate vectors provide both native promoter constructs and fusions driven by the strong cauliflower mosaic virus 35S promoter (*see* **Note 2**), with GFP and derivates at both the 5′- and 3′-end of the coding sequence of interest. Note that these fusions are more likely to have targeting or folding defects than genes cloned via FTFLP with internal GFP protein tags.

GFP is an outstanding marker for localization of a single tagged protein, but it is not particularly well-suited for investigation of protein-protein interactions, due to an overlap in emission spectra with its most robust spectral variants, yellow- and cyan-fluorescent proteins (YFP and CFP, respectively). The best spectral separation from GFP is in the red part of the spectrum, but we have not used the GFP-like proteins from tropical corals (*Discosoma* sp.) that emit in the orange-red region (DsRed), because of interference from chlorophyll autofluorescence, reduced fluorescence lifetime, and problematic formation of DsRed multimers in vivo *(11)*. However, recent improvements of GFP-derived modified red-fluorescent protein (mRFP) have led to very bright, stable chromophores which do not form multimeric complexes in vivo *(12)*. Although the new mRFP shows great promise, we have had success with simultaneous colocalization of Citrine YFP and CFP and continue to use these fluorophores. The CFP can be excited at 440 nm, which is too short a wavelength to excite YFP. An excitation wavelength of 514 nm will excite YFP, but cannot stimulate CFP. Accordingly, emission detection at 450–490 and 525–600 nm can be used without overlap [see *(4)* for a review of the applications of fluorescent proteins, including colocalization studies].

Multifluorescent plants have uses beyond simple co-localization, but many of these techniques are very involved and do not lend themselves to high-throughput imaging. However, improvements in techniques such as fluorescence resonance energy transfer (FRET) may eventually lead to increased capabilities for high-throughput protein-protein interaction assays. Fluorescence lifetime imaging microscopy (FLIM) is a promising new technique for resolving different fluorophores, which relies on the decay time rather than color to differentiate between FRET and normal emission (*see* **Note 3**).

In this chapter, we present a start-to-finish approach to high-throughput imaging of Arabidopsis that caters to two different audiences: those who are working with plants for the first time, and plant scientists looking to apply high-throughput imaging to existing projects. The foremost web resource for the Arabidopsis community is The Arabidopsis Information Resource (TAIR; **Table 1**). Through TAIR, researchers can search genomic resources and find seed stocks for various marker lines, characterized mutants, and uncharacterized insertion lines for reverse genetics (*see* also 1).

2. Materials

2.1. Storage and Sterilization

1. Small coin envelopes (Office Max; Naperville, IL).
2. 0.1% Gelrite (Research Products International; Mt. Prospect, IL).
3. Ethanol sterilization solution: 95% ethanol, 0.1% Triton-X 100.
4. Bleach sterilization solution: 50% bleach, 0.1% Triton-X 100; made fresh each day to avoid detergent precipitation.
5. Ultra-pure water, autoclaved and deionized.

2.2. Plate Preparation

2.2.1. Petri Plates

1. Petri plate.
2. Arabidopsis growth media (AGM): Murashige & Skoog salts for 1 L (bioWORLD; Dublin, OH), 20-g sucrose, ultra-pure water to 2 L (making the final mixture half-strength). Add 750-mg Gelrite, or 1,500 mg of any appropriate agar, to each 250-mL aliquot of liquid AGM. Autoclave 30 min. Sterile AGM can be stored in bottles at room temperature for up to 6 months if no contamination is evident, or on plates for up to one month. To use, microwave solid AGM for 5 min at a time on low power until completely melted. Carefully monitor and shake the bottle so that the media does not boil over. Cool the AGM in a 65°C water bath until the bottle is cool enough to handle. Use sterile technique at all times when handling and aliquotting AGM—bacteria and fungi thrive in this sucrose-based media.
3. Standard microscope slide and 24 mm × 50 mm cover glass.
4. Paper surgical tape.

2.2.2. Rectangular Universal Lids

1. Standard "universal" lid, 128 mm × 86 mm, no thatched corner (Corning Costar; Lowell, MA).
2. 120 × 75 #1 cover-glasses (Erie Scientific; Portsmouth, NH).
3. Arabidopsis growth media (AGM; *see* **Subheading 2.2.1**).

2.2.3. Forty-Eight-Well Chemical Plates

1. Standard "universal" lid, 128 mm × 86 mm, no thatched corner (Corning Costar).
2. Gasket sets (Grace Bio-labs; Bend, OR).
3. 120 mm × 75 mm # 11/2 cover glasses (Erie Scientific).
4. Arabidopsis growth medium (AGM; *see* **Subheading 2.2.1**)

3. Methods

3.1. Storage and Sterilization

1. Due to their small size, huge numbers of Arabidopsis seeds can be stored in limited space; the most economical method of storage is in small coin envelopes.

2. These envelopes can be stored upright in boxes or bags and should survive 2 years at room temperature (some mutants have decreased viability over time).
3. To prolong seed life, store seeds in airtight containers at 4°C with fresh silica gel desiccant as needed.
4. It is advisable to bulk a fresh stock and store at least 1,000 seeds from any new genotype in long-term refrigerated storage in dessicant. If possible, store permanent stocks away from the main lab area to protect against fire or vandalism. These seeds should be re-planted and bulked approximately every 2 years.

3.1.1. Seed Bulking

This procedure is used to propagate seeds for storage and future use in large quantities. After bulking, it is advisable to store at least 1,000 seeds in a 1.5-mL tube or small coin envelope at 4°C in a desiccated environment.

1. Examine plants phenotypically or genotype seedlings for transgenes or mutations to ensure that these traits are passed to progeny.
2. Transplant 10–20 genotyped individuals to soil.
3. If a population of dry seeds has been adequately verified, suspend several hundred vernalized seeds in 0.1% Gelrite or other agar in a 15-mL conical tube.
4. Shake and vortex the tube vigorously to separate and suspend seeds.
5. Use a transfer pipette or 1-mL micropipettor to distribute suspended seeds evenly across the soil surface.
6. Once siliques begin to dry and shatter, discontinue watering until plants are dry, and harvest seeds by gently tapping plants over a white piece of paper.
7. Shake seeds back and forth across the paper so that light chaff and plant material are easier to remove with fingers, tweezers, or by gently blowing debris off the paper. Take care to contain and autoclave all transgenic material.

3.1.2. Seed Sterilization

Since plant growth medium is also an ideal formula for the growth of bacteria and fungus, seeds must be thoroughly sterilized before planting. It is best to first use ethanol to eliminate fungal contaminants, followed by a bleach/detergent solution to kill bacteria and remaining fungus completely. In most cases, bleach alone is sufficient, but fungus can quickly ruin an experiment, so add the preliminary sterilization step if fungal contamination becomes problematic.

1. Sterilize all surfaces in a laminar flow hood with 70% ethanol and ensure that air flows unimpeded to the work area.

2. Aliquot the desired number of seeds (1 g = 50,000) into a 1.5-mL micro-centrifuge tube.
3. (optional) Add 1-mL ethanol sterilization solution to seeds. Shake the tube and rock or rotate for 5 min. Remove the ethanol and rinse seeds twice with sterile water.
4. Add 1-mL bleach sterilization solution to seeds.
5. Shake vigorously to eliminate seed clumps and rock or rotate for no more than 7 min. The bleach solution will begin to kill seeds after 10 min of exposure. Do not sterilize more than 20 tubes at a time so that seeds are not overexposed while you work. Seeds that have turned white in bleach solution are overexposed and will not germinate.
6. Centrifuge briefly at 7,000 rpm to settle seeds and clear bleach from tube lids.
7. Remove bleach solution using either a 1-mL pipette or a vacuum trap device with sterile 200-μL tips (*see* **Note 4**).
8. Add 1-mL ultra-pure water to each tube, taking care to dislodge clumps from the bottom for thorough rinsing.
9. Allow seeds to settle, and repeat steps 4–5 for a total of four changes of water to ensure that no bleach remains. Leave 1-mL water in each tube.
10. Seeds should then be cold-treated (vernalized) in sterile water to ensure even germination once they return to a warm environment. This can be done by placing tubes (each containing seeds and 1-mL sterile water) in a dark box at 4°C for 3–7 days.

3.2. Plate Preparation

Choosing a plate design that suits a particular experiment is critical for maximizing throughput. Thus, we sought to eliminate the most time-consuming, damaging step in traditional microscopy: transferring seedlings from growth media to microscope slides. Transfer time is minimal when a few seedlings are examined, but becomes significant when screening very large numbers of seedlings in applications such as map-based cloning or chemical genomics. To this end, we have developed two unique systems for high-throughput confocal imaging of Arabidopsis seedlings that practically eliminate plant handling during image collection.

3.2.1. Petri Plates

The traditional method for microscopic examination of Arabidopsis seedlings involves growth plants in Petri dishes (**Fig. 2A**), followed by transfer to microscope slides with cover glasses. This technique is relatively low-throughput and can cause tissue damage, but is discussed briefly here as a reference point for other higher-throughput methods.

1. Prepare Arabidopsis growth media.
2. Using sterile technique in a laminar flow hood, pipette or pour enough liquid media to cover the bottom of the plate to

a depth of ~5 mm (30 mL for 3.5 in. round, 90 mL for 6 in. round, 40 mL for 3.5 in. square).

3. Place sterile vernalized seeds onto agar surface (*see* **Subheading 3.3.1 and 3.3.2**).
4. Seal plates with paper surgical tape. Do not use Parafilm or non-porous tape, which will prevent gas exchange necessary for photosynthesis.
5. Grow for 7–10 days in a growth chamber (*see* **Subheading 3.4**) before imaging.
6. To prepare for imaging, remove seedlings from agar media with sterile fine forceps and carefully lay them onto a glass slide.
7. Add a few drops of water and cover the seedlings with a 24 mm × 50 mm #1 1/2 cover glass *see* **Subheading 3.5.2**).

3.2.2. Rectangular Universal Lids

1. Prepare Arabidopsis growth media.
2. Using sterile technique in a laminar flow hood, pour just enough liquid media to cover the bottom of the plate, approximately 20 mL.
3. Place sterile seeds onto agar surface (*see* **Subheading 3.3.1** and **3.3.2**).
4. Assemble pairs of plates so that seedlings face inward, maintaining a sterile environment inside. To increase storage density, additional plates can be stacked facing inward onto the backs of the original pair for a total of up to five plates.
5. Seal plates with paper surgical tape. Do not use Parafilm or non-porous tape, which will prevent gas exchange necessary for photosynthesis.
6. Place plates vertically in a growth chamber (*see* **Subheading 3.4.**), and use racks or other devices as necessary to maintain vertical orientation. Roots will grow downward along the agar surface.
7. Grow seedlings for 2–10 days in a 22°C growth chamber on a 16/8-h light/dark cycle.
8. To prepare for imaging, remove tape and separate plates. Add a small amount of sterile water to the agar surface, taking care not to disturb the seedlings.
9. Gently press the large cover-glass onto one end of the plate and gradually lower the glass onto the agar surface over the seedlings. Add water and/or tap gently as necessary to eliminate bubbles.
10. Flip the plate quickly and allow excess water to drain onto a paper towel.

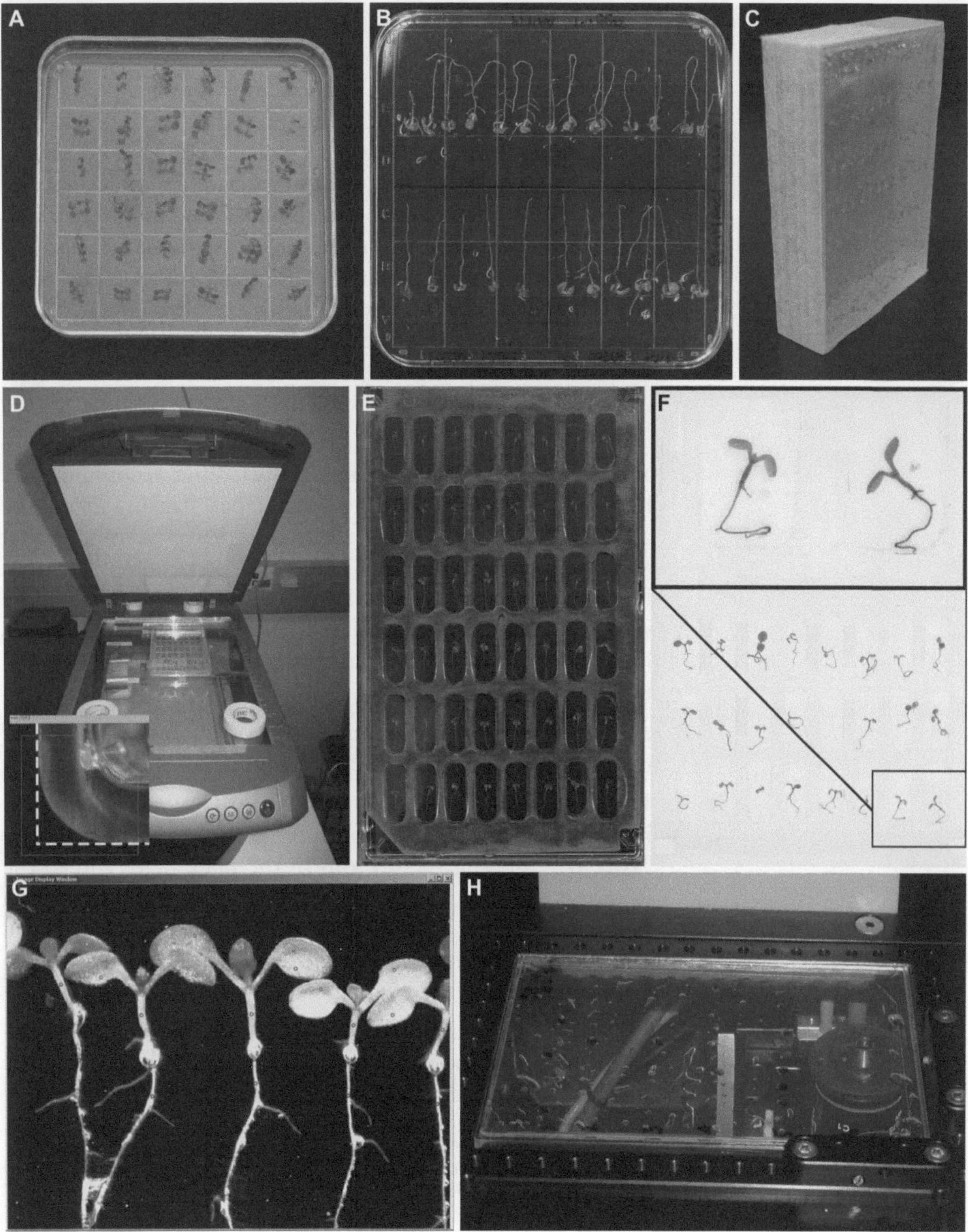

Fig. 2. (A) Square petri plates for growing Aribidopsis vertically or horizontally. (B) When used vertically, the plate pictured in (a) can be rotated to test gravitropism. In this example, response to salt stress is revealed by an absence of new root growth in a downward direction. (C) Universal plate lids covered with a thin skin of agar growth medium can be stacked together and sealed with paper tape for high-density seedling incubation. (D) The middle of a conventional flatbed document scanner gives excellent images of plated seedlings, which can be used as a reference map for semi-automated microscopy. Mapping accuracy depends on exactly matching the reference corner of the plastic plate to the scan area marquee (insert). (E) A silicone rubber gasket insert converts a universal lid into 48 wells for chemical screens. One layer of the gasket set is replaced with a large cover glass for mapping and microscopy of all tissues without the need to handle seedlings. (F) Fluorescence images from a gel scanner can also be used for mapping plates, with the advantage that GFP expression levels are recorded directly. Fluorescence appears dark on a white background. (G) The user annotates the map image with hit sites, which are converted by Point Picker into vector files for visiting those locations with the automated microscope. (H) A plate being scanned. Hit positions can be recorded with pen marks on the back of the plate.

3.2.3. Forty-Eight-Well Chemical Plates

Although this design requires a bit more time for plate preparation and greater expense than the other methods discussed, the 48-well plate setup allows for an isolated well for each seedling (**Fig. 2E, F**). This setup is particularly applicable to chemical-genomics screens, where users study the effects of thousands of different chemicals on experimental subjects such as Arabidopsis. In addition to the universal lids and large rectangular cover glasses used in the previous section, this design requires a silicone rubber gasket (Grace Biolabs, Berd, OR) set to create isolated wells for seedlings.

1. Prepare Arabidopsis growth media, but instead of cooling in a water bath as discussed previously, the agar must be very hot.
2. Prepare 48-well gasket plates using sterile technique in a laminar flow hood. Remove the paper backing from the adhesive side of the gasket set, then lower it middle-first onto a universal plate lid. There is a central hole, which can be lined up with the dimple of the plate center. Once the middle touches, make sure it is square, then lower both ends into the plate. Press the gasket all the way to the top and bottom edges (*see* **Note 5**).
3. Aliquot appropriate concentrations of library chemicals or drugs into wells using a robot or multichannel pipettor.
4. Using sterile technique, pipette 200 mL of very hot AGM into each chamber with a robot or multichannel pipettor (*see* **Note 6**). Tapping the bottom of the plate helps spread the liquid to wet all sides of each well uniformly. A plastic trough of freshly boiled agar will remain hot enough for several plates, but if it cools too much, the pipette tips will begin to clog and the media will not spread easily into each well.
5. Place a single sterile seed onto the agar surface (*see* **Subheading 3.3.1** and **3.3.2**) approximately **1/3 of the way from the top, to allow room for root growth.
6. Growth is improved by pushing a couple of crystals of ethylene absorber (*see* **Note 7**) between the gasket and the top rim of the plate.
7. Assemble pairs of plates so that seedlings face inward, maintaining a sterile environment inside. To increase storage density, additional plates can be stacked facing inward onto the backs of the original pair for a total of up to five plates (**Fig. 2C**).
8. Seal plates with paper surgical tape. Do not use Parafilm or non-porous tape, which will prevent gas exchange necessary for photosynthesis.
9. Place plates vertically in a 22°C growth chamber (*see* **Subheading 3.4**), and use racks or other devices as necessary

to maintain vertical orientation. Plants will grow within the chamber on top of the agar surface with roots growing downward. Growth of longer than 7 days will exceed the chamber capacity.

10. To prepare for imaging, separate taped plates and peel back the top layer of the silicone gasket. Take care that no roots are overhanging the agar pads, where they may be pulled off with the gasket.
11. Add a small amount of sterile water to the agar surface, taking care not to disturb the seedlings.
12. Gently press the large cover glass onto one end of the plate and gradually lay the glass onto the agar surface over the seedlings. Add water and tap gently as necessary to eliminate bubbles.
13. Quickly flip the plate over and remove excess water with a paper towel.

3.3. Plating Seeds

3.3.1. Plating Sterile Seeds with a Microchannel Pipettor

Arabidopsis seeds are very small and thus single seeds cannot be placed by hand or with tweezers. There are many seed placement methods, but this technique allows for very fast and precise dispensation of Arabidopsis seed with minimal practice.

1. Sterile seeds (in water) are taken up into a 1-mL pipette, approximately 5–10 seeds at a time, in ~100-μL water.
2. The pipette should be set to maximum volume (1 mL), but the user should only use the last 10% of the volume range, so that the hand can relax on the pipette and not depress the plunger while working.
3. The seeds will slowly settle and line up at the tip of the pipette. They can be dispensed one at a time by gently touching the pipette tip to the surface of the agar. This technique is accomplished by using surface tension between the agar media and the water surrounding the seed in the pipette tip, not by plunging with the pipette (which is the intuitive method for most beginners). The key is that pressure on the plunger is not necessary, and will cause more than one seed at a time to leave the tip.
4. If multiple seeds are consistently dispensed at once as the tip touches the agar, reduce the number of seeds in the tip to 3–5.
5. Once the technique becomes more comfortable, increase the number of seeds to 10–20, and re-dip the tip into sterile water every few times the agar is touched. This replenishes the water in the tip and allows good contact between the seeds and the agar for easy dispensing.

3.3.2. Automated Single Seed Placement

We have also developed a successful prototype seed-placement approach, which lends itself to automation.

1. A blunt ended 26-gauge needle (Small Parts Inc, Miramas, FL) will pick up a single seed when a vacuum is applied to it.
2. Releasing the vacuum leaves the seed held in place by surface tension, but when it is touched against an agar surface, the seed transfers to an exact position.
3. Rapid dip and place can be used to populate plates, using a vacuum pick-up system, modified to include a small water trap in the vacuum line.
4. Keller-Swartwood Engineering (Aurora, Oregon) developed this seed handling approach for us and we have a prototype pneumatic seed-placement robot head on a peg-board test rig. The head swings from vertical for pick-up to a 45° angle for dispensing onto an agar surface. The inclined angle makes vertical precision less important, since the needle flexes as it touches the agar and displaces the seed without burying it below the surface or forcing it back down the needle aperture. It has 100% pick-up reliability and placement accuracy to within one seed diameter.

3.4. Growth Chamber Incubation

1. Seedlings grown vertically under even light will grow uniformly on the surface of the agar growth media.
2. A CU-36L4 incubator (Percival; Perry, IA) has four lighted shelves, with sufficient space between for vertical high-throughput plate lids.
3. The more common five-shelf growth chamber brings the upper row of seedlings too close to the fluorescent tubes, which then have to be shaded with a loose sheet of cheesecloth.

3.5. Imaging

3.5.1. Choice of Instrument and Objective Lens

There are many microscopy systems available for high-throughput screening, but most are designed for observation of adherent animal cells. We selected the BD Pathway HT system (BD Biosciences; Rockville, MD) as our primary instrument, because it is highly automated, camera-based, and has a spinning-disc confocal capability, which can be pulled out of the light path for conventional fluorescence imaging. It moves one objective lens around on linear motors under a stationary sample, which makes it uniquely amenable to non-adherent plant cell cultures, because the cells are not disturbed by movement of the plate.

We have performed mutant screens on an upright Meridian InSight ocular viewing confocal microscope, using hand-eye coordination for rapid navigation across slide-mounted seedlings. This company no longer exists, but the BD Bioscience CARV 2 confocal imager is its closest modern equivalent, and can be used with inverted motorized stands. An upright stage can be used with high-throughput plates, but only half of the plate is accessible at a time. We also have a high-sensitivity Yokogawa CSU 10 confocal system, with an Optical Insights

DualView eyepiece, which can image CFP on one side of the camera and simultaneously capture YFP signal on the other.

High numerical aperture (NA) objectives give the best confocal optical sectioning, and brightest fluorescence signal. However, bulky lenses are unable to approach the edge lip of a plate lid, eliminating two entire rows and columns from each plate. Olympus has a pair of very good 20×/0.75NA and 40×/0.9NA UApo340 lenses with tapered front ends, which can address all wells in a 96- or 384-well footprint. On an upright stand, any dipping objective should work fine. Ultra-long working distance lenses are also possibilities, but they tend to have much lower numerical apertures. For follow-up work, water-immersion lenses have the best combination of NA and working distance for plant work. The higher NA of oil objectives gives no real advantage in live samples, unless the point of interest in the sample is pressed against the cover glass.

3.5.2. Observation Under Cover Glasses

1. Additional 18 mm × 18 mm cover glasses can be used to make a bridge around larger seedlings which have their first true leaves.
2. As long as there is some air below the span, surface tension will pull down on the larger cover glass and prevent it from floating around.
3. Alternatively, strips of petroleum jelly extruded through a thick hypodermic needle can be used to mount a row of up to 10 seedlings under a 24 mm × 60 mm cover glass (*see* **Note 8**).

3.5.3. Semi-automated Imaging

1. We used an Epson Perfection 4180 Photo document scanner.
2. Scanning needs to be exactly parallel to the sides of the plate, so that distances from the reference corner are predictable enough to give a good placement of the hit site.
3. We found that the middle of the document area gives better results than one side, because it avoids the parallax of the scanner's light path hitting a surface that is a bit above the glass surface.
4. Two sheets of knife-maker glass worked well as a spacer to bring the plate close to the middle.
5. To correlate scanner positions to those in the microscope, we made a reference plate containing EM grids. These are large enough for point-picking on the map image, and have a central check-mark, which precisely locates the grid center, for measuring the acquired position on the automated microscope.

3.5.4. Semi-automated Imaging on a Gel Scanner

1. The Typhoon 9400 system (Amersham Bioscience, Pittsburgh, PA) has various fluorescence settings for different sample types. It also allows scanning at 3 mm above the surface of the glass pattern.
2. Imaging can be performed at any resolution from 10 to 1,000 μm, at a cost of speed.

3. We endeavored to keep the mapping and imaging process to under an hour, to minimize any stress effects on the seedling while it is under glass.
4. Use 200-μm pixels to preview and assure the image has enough brightness, but is not saturated, and then switch to either 25-μm, which takes 25 min, or 50-μm, which takes 13 min. Our Point Picker software has different calibrations for each scanner and scanner setting. We set up a separate calibration for each of the 48 wells, but this did not improve precision, which is mostly determined by care in clicking on positions of interest, and care in moving from scanner to microscope.

3.5.5. Automated Imaging

It is surprisingly difficult to develop software to detect features of interest in complex samples automatically, since software is easily distracted by irrelevant features like shadows, air bubbles, condensation, or scratches on the plate. However, a semi-automated approach, where the user clicks on a map of the sample, has proven to be quite effective in speeding up the imaging process.

1. Place the coverslipped plate lid sample on a calibrated flatbed scanner (*see* **Subheading 3.5.3**; **Fig. 2D**). Leave the lid open for a dark background and maximum contrast, to highlight light-colored plant tissues such as roots.
2. Take a preview scan, and then draw two marquees; one to cover the whole plate, and a second to cover a very small area around the reference corner of the plate.
3. Take a zoom image of the small marquee, and use this enlarged image to place the big marquee rectangle over the very edge of the plastic plate exactly (**Fig. 2D**, insert). The Epson software allows these marquees to be retained from one use to another, but they can easily be redrawn.
4. Switch to the big rectangle and make a high-resolution scan at 1,200 dpi.
5. A fluorescence plate imager, such as the Typhoon 9400 (Amersham Bioscience, Pittsburgh, PA) can be used instead of a flatbed scanner to give a map of fluorescence signal rather than just morphology (**Fig. 2F**). This may be useful in mutant screens where only a percentage of seedlings are correctly labeled, and the rest can be discounted at the map stage (*see* **Subheading 3.5.4**).
6. Point-picking software can now be used to annotate this map image with "hit" sites, and generate a series of visit coordinates for an automated microscope. We initially adapted Image J to perform this function, before writing our own software in Python for generating these tables (*see* **Subheading 3.5.6**).
7. Transfer the plate to the automated microscope, taking care to match the position of the reference corner with the reference corner of the plate holder in the imager.

8. Each "hit" point should be accessible to the microscope at a precision of <50 μm. You can select a location, and then navigate from the "hit" location to evaluate the tissue more thoroughly. Alternatively, a macro can be used to auto-focus, and snap an image at all locations in sequence (*see* **Subheading 3.5.7**).

3.5.6. Point-Picking with Image Software

Our Point Picker software is available for download at: http://bioweb.ucr.edu/ChemMineV2/protocols. Alternatively, Image J is a freeware program, with a vibrant community of users who supply modules and plugins for a wide range of uses. We used *plugin_to_handle_extra_file_types, specify_ROI, Time_Stamper, and Point_picker* to enable us to convert the table generated by a series of clicks on the map image, so that it could feed our BD Pathway HT imager with the right format of x-y-z vector files.

3.5.7. Auto-focus

For seedlings, the auto-focus feature must be image-based, with the microscope moving through a range of focus levels and selecting the height with the greatest brightness and contrast. Laser-based auto-focus bounces a beam off the cover glass, and then offsets a given distance, which is faster and more effective for cultured animal cells, but will not work on three-dimensional structures such as whole seedlings. Cameras can also auto-expose, to give an adequate image, even when some regions are very brightly stained and others quite dim.

4. Notes

1. pEarlyGate Vectors are available at: http://www.biology.wustl.edu/pikaard/pEarleyGate%20plasmid%20vectors/pEarleyGate%20homepage.html.
2. The strong "constitutive" 35S promoter, although highly active in most tissues, has very little expression in Arabidopsis pollen. Alternatively, the LAT52 promoter has enhanced expression in mature pollen and throughout germination *(13)*.
3. Fluorescence Lifetime Imaging Microscopy (FLIM) is explained (http://www.picoquant.com/_scientific.htm).
4. With minor customization, a side-arm flask is a convenient device for rapidly washing sterilized Arabidopsis seeds. In a sterile hood, attach a vacuum line to a stopper on the top of the flask, and run an additional hose from the side arm. A trimmed Pasteur pipette on the end of the side-arm vacuum line makes a convenient handle for sterile 200-μL tips, and can maintain a sterile environment for the seeds. For each sample, attach a new tip to the end of the Pasteur pipette and remove liquid under a gentle vacuum, taking care to leave seeds undisturbed at the bottom of the tube.

5. A trimmed soft rubber print-roller works well for preparing 48-well plates (Electron Microscopy Science; Hatfield, PA).
6. Robotic dispensing of AGM: A BioTek Precision 2000 8-tip robot (BioTek; Winooski, VT) is sufficient for preparing assay plates with solid media. The instrument is simple and reliable, and is compact enough to be placed in a laminar flow hood for aseptic setup. We also use a large and more troublesome Biomek FX^P robot (Beckman Coulter; Fullerton, CA), to which we have added a vinyl curtain and HEPA air blower (IQAir; Santa Fe Springs, CA) for sterility, and mounted a hot-plate stirrer (Corning) on the robot deck for maintaining agar.
7. Extend-a-Life Ethylene Absorber (Agraco; Normstown Hill, PA) consists of sachets of zeolite granules containing dry potassium permanganate and is used to preserve fruits, vegetables, and flower arrangements by removing ethylene, given off during normal ripening. It works to improve the uniformity and growth rate of Arabidopsis seedlings. This material does not work when wet, so push a couple of granules into the top corner of a 48-well gasket plate, where condensation will not drip down and spoil its activity.
8. Erie Scientific, (Portsmouth, NH) makes cover glasses and slides to almost any specification, and with less than 1-month lead time, provided they are ordered in bulk.

Acknowledgments

We wish to acknowledge the vision and leadership of Natasha Raikhel, who is director of the Center for Plant Cell Biology (CEPCEB), the Institute for Integrative Genome Biology (IIGB), and was instrumental in bringing chemical-genomics capability to UC-Riverside. Matt Mason developed the ability to screen plate lids on the Pathway HT, and Anh Vu did all the coding for the Point Picker software.

References

1. Weigel, D. and Glazebrook, J. (eds.) (2002) *Arabidopsis: A Laboratory Manual.* Cold Spring Harbor Press, Cold Spring Harbor, NY.
2. Avila, E. L., Zouhar, J., Agee, A. E., Carter, D. G., Chary, S. N., and Raikhel, N. V. (2003) Tools to study plant organelle biogenesis. Point mutation lines with disrupted vacuoles and high-speed confocal screening of green fluorescent protein-tagged organelles. *Plant Physiol.* 133, 1673–1676.
3. Chary, S. N., Robert, S., Yang, Z., Raikhel, N. V., and Hicks, G. H. (2007) Chemical genomics of networks controlling vesicular trafficking in plant development. http://bioweb.ucr.edu/ChemMineV2/twentyten.
4. Brandizzi, F., Fricker, M., and Hawes, C. (2002) A greener world: The revolution in

plant bioimaging. *Nat. Rev. Mol. Cell Biol.* 3, 520–530.

5. Shimomura, O., Johnson, F. H., and Saiga, Y. (1962) Extraction, purification and properties of aequorin, a bioluminescent protein from the luminous hydromedusan. *Aequorea. J. Cell. Comp. Physiol.* 59, 223–239.
6. Brandizzi, F., Irons, S. L., Johansen, J., Kotzer, A., and Neumann, U. (2004) GFP is the way to glow: Bioimaging of the plant endomembrane system. *J. Microsc.* 214, 138–158.
7. Tian, G. W., Mohanty, A., Chary, S. N., Li, S., Paap, B., Drakakaki, G., Kopec, C. D., Li, J., Ehrhardt, D., Jackson, D., Rhee, S. Y., Raikhel, N. V., and Citovsky, V. (2004) High-throughput fluorescent tagging of full-length *Arabidopsis* gene products in planta. *Plant Physiol.* 135, 25–38.
8. Cutler, S. R., Ehrhardt, D. W., Griffitts, J. S., and Somerville, C. R. (2000) Random GFP: cDNA fusions enable visualization of subcellular structures in cells of *Arabidopsis* at a high frequency. *Proc. Natl. Acad. Sci. USA* 97, 3718–3723.
9. Li, S., Ehrhardt, D. W., and Rhee, S. Y. (2006) Systematic analysis of Arabidopsis organelles and a protein localization database for facilitating fluorescent tagging of full-length Arabidopsis proteins. *Plant Physiol.* 141, 527–539.
10. Earley, K. W., Haag, J. R., Pontes, O., Opper, K., Juehne, T., Song, K., and Pikaard, C. S. (2006) Gateway-compatible vectors for plant functional genomics and proteomics. *Plant J.* 45, 616–629.
11. Jakobs, S., Subramaniam, V., Schonle, A., Jovin, T. M., and Hell, S. W. (2000) EGFP and DsRed expressing cultures of Escherichia coli imaged by confocal, two-photon and fluorescence lifetime microscopy. *FEBS Lett.* 479, 131–135.
12. Jach, G., Pesch, M., Richter, K., Frings, S., and Uhrig, J. F. (2006) An improved mRFP1 adds red to bimolecular fluorescence complementation. *Nat. Methods* 3, 597–600.
13. Eady, C., Lindsey, K., and Twell, D. (1993) Differential activation and conserved vegetative cell-specific activity of a late pollen promoter in species with bicellular and tricellular pollen. *Plant J.* 5, 543–550.

Chapter 7

Fluorescence-Based Assays

W. Frank An

Summary

Fluorescence-based assays are widely used in high-throughput screening due to their high sensitivity, diverse selection of fluorophores, ease of operation, and various readout modes. As a result, fluorescence-based assays have been applied to monitor a broad range of activities in life-science research such as molecular dynamics and interactions, enzymatic activities, signal transduction, cell health, and distribution of molecules, organelles, or cells. This chapter describes two fluorescence-based techniques: total intensity measurement as an indication of cell viability, and fluorescence resonance energy transfer as an indication of protein folding and interactions, to illustrate in detail applications suitable for cell-based high-throughput screening in plate-reader and automated microscope-based formats, respectively.

Key words: Automated microscope, Calcein, Cell-based, Fluorescence, Fluorescence resonance energy transfer, High-content image-based screen, High-throughput screen, Nuclear translocation.

1. Introduction

Fluorescence-based detection is arguably the most dominant applied detection method in life-science research, particularly in applications that involve high-throughput screens (HTS). This widespread use is largely due to very high sensitivity, diverse selection of fluorophores that excite and emit across a broad spectrum of wavelengths, and variety of types of readouts based on different environment-sensitive fluorescence properties. These unique characteristics of fluorescence-based detection techniques allow miniaturization, flexibility in assay design, ease of operation, and simultaneous monitoring of multiple events, thus making fluorescence-based assays quite amenable to HTS.

P.A. Clemons et al. (eds.), *Cell-Based Assays for High-Throughput Screening, Methods in Molecular Biology, Vol. 486*

DOI: 10.1007/978-1-60327-545-3_1

Fluorescence-based assays can be generally divided into two classes. The first class encompasses techniques that macroscopically detect the total fluorescence intensity, fluorescence polarization, fluorescence resonance energy transfer (FRET), fluorescence lifetime, time-resolved fluorescence, and combinations of these techniques, such as time-resolved fluorescence polarization. The second class of fluorescence-based assays detects fluorescence from single fluorescent molecules, such as fluorescence correlation spectroscopy and fluorescence intensity distribution analysis. These fluorescence techniques have been used to monitor an enormous collection of biological processes, such as macromolecule–macromolecule interactions, macromolecule–small molecule interactions, enzymatic activities, signal transduction, cell health, and states and locations of molecules, organelles, or cells [for general information on fluorescence and its applications, *see (1–3)]*. Given the tremendous scope of fluorescence-based applications, as well as the emphasis of this textbook on cell-based high-throughput assays, this chapter will not endeavor to cover all these techniques and applications. Instead, it will focus on two commonly used techniques: a total fluorescence intensity measurement using a fluorescent dye, calcein, as an indicator for cell viability, and FRET measurement in an image-based format using cyan-fluorescent protein (CFP) and yellow-fluorescent protein (YFP) as an indicator for intramolecular folding and intermolecular association of proteins. The following protocols apply to batch processing with individual instruments, with emphasis on the conditions and procedures that impact on assay biology, as opposed to certain other important operational aspects of HTS, such as compound management and informatics.

2. Materials

2.1. Calcein-AM Viability/Cytotoxicity Assay

2.1.1. Assay Background

Calcein-AM is the acetoxymethyl ester (AM) derivative of calcein. There are two significant differences between calcein-AM and calcein. Calcein-AM is colorless and nonfluorescent until hydrolyzed into calcein. Calcein-AM is cell membrane-permeable and can readily enter cells by diffusion. In contrast, calcein is fluorescent and, owing to its six negative charges and two positive charges at neutral pH, is cell membrane-impermeable *(4)*. Therefore, once calcein-AM enters cells with an intact membrane, the AM-ester moiety will be hydrolyzed by nonspecific esterases in live cells and the resulting product, calcein, will be trapped inside such cells, generating fluorescence upon excitation. Calcein's superior cellular retention over other popular fluorophores, such as fluorescein and BCECF, bright fluorescence insensitive to physiological pH, and lack of interference with several cellular processes make it, in its AM-ester form, a widely used fluorescent probe to mark live cells with an intact cytoplasmic membrane *(4)*. Specifically, calcein-AM has been used to measure cell viability and cytotoxicity (*see* **Note 1**), membrane permeability, gap-junction formation, cell adhesion, and multidrug resistance.

2.1.2. Consumables

1. *Cell culture media.* The following protocol applies to virtually all adherent cell types. Because of this, no cell lines and growth conditions are specified here, but most cells grow with their base media supplemented with 5–10% fetal bovine serum (FBS; HyClone, Ogden, UT). Inclusion of antibiotics in media is encouraged, as long as the antibiotics of choice (e.g., penicillin and streptomycin; Invitrogen, Carlsbad, CA) are compatible with the cell line-specific culture conditions. This protocol should apply independently of common medium types and FBS percentages.
2. Trypsin-EDTA (Invitrogen).
3. Phosphate-buffered saline (PBS; Invitrogen).
4. Calcein-AM (Invitrogen).
5. 384-Well black, clear-bottomed plates (*see* **Note 2**).

2.1.3. Instrumentation

1. Fluorescent plate reader with top-reading capability. Many brand names will work; however, those with stackers, such as the EnVision (PerkinElmer, Waltham, MA) and Synergy (BioTek, Winooski, VT), provide much better throughput.
2. Bulk dispenser with plate stackers, such as μFill (BioTek), WellMate (Matrix Technologies, Hudson, NH), and the Multidrop (Thermo Scientific, Waltham, MA).
3. Plate washer with plate stackers, such as ELx 405 (BioTek).
4. Compound-transfer instrument that is able to deliver nanoliter (nL) range of com1pounds, such as the CyBi-Well (CyBio, Woburn, MA) equipped with a 384-pin head.
5. Tissue culture incubators.

2.2. CFP-YFP-Based FRET Measurement in Conjunction with Fluorescence-Based Nuclear Translocation Monitoring in an Image-Based, High-Content Assay

2.2.1 Assay Background

FRET is a distance-dependent energy transfer mechanism, by which the excited state of the donor fluorescent molecule is transferred to a second, acceptor fluorescent molecule without fluorescence radiation from the donor *(1)*. FRET requires that the emission spectrum of the donor fluorophore overlaps with the excitation spectrum of the acceptor fluorophore, and that these two fluorophores are in close proximity (10–100 Å) *(1)*. The distance-dependent property of FRET is employed extensively in bioscience to monitor complex formation or dissociation, or any activity that may cause complex formation or dissociation, such as ligand binding-induced conformational changes *(4)*. CFP (as donor) and YFP (as acceptor) constitute a pair of fluorophores that have been widely used as FRET-generating partners (*see* **Note 3**). This is because the emission spectrum of CFP overlaps with the excitation spectrum of YFP. Therefore, CFP-to-YFP FRET takes place when respective CFP- and YFP-tagged macromolecules, such as proteins, peptides, or DNA sequences, are brought sufficiently close to each other through a biological mechanism, such as intramolecular folding, protein–protein, protein–DNA, and DNA–DNA interactions, or ligand binding.

The following is an outline of an image-based application of CFP-YFP FRET to identify small-molecule modulators of intramolecular folding and/or dimerization of androgen receptor (AR) using a CFP-AR-YFP fusion protein (C-AR-Y) expressed in HEK293 cells. The application uses an automated epifluorescence microscope that is suitable for high-content screening. In addition to intramolecular folding/dimerization of AR indicated by increased FRET signals upon stimulation of dihydrotestosterone (DHT, ligand for AR), the high-content format of this image-based application allows the simultaneous monitoring of another critical feature of AR function, AR nuclear translocation, by tracking cellular location of each fluorescence tag (*see* **Note 4**).

2.2.2. Consumables

1. Cell lines and cell culture conditions.
   a. HEK293 cells (*see* **Note 5**).
   b. An HEK293 cell line stably transfected with a constitutively expressing C-AR-Y construct (*see* **Note 5**). This line is hereafter referred to as C-AR-Y/HEK.
   c. *Culture media*. For HEK293: DMEM (Invitrogen) supplemented with 5% FBS and 1× penicillin/streptomycin mixture. For C-AR-Y/HEK: the same medium for HEK293, further supplemented with 5 μg/mL hygromycin (Invitrogen).
2. Trypsin-EDTA (Invitrogen).
3. DHT (Sigma-Aldrich, St. Louis, MO).
4. Formaldehyde, 37% (Sigma-Aldrich).
5. Hoechst 33342 (10 mg/mL, Invitrogen).
6. Phosphate-buffered saline.
7. 384-well black, clear-bottomed plates (Corning Costar, Lowell, MA; *see* **Note 6**).
8. Opaque plate seals (PerkinElmer).
9. Fix solution: 4% formaldehyde and 0.5 mg/mL Hoechst in PBS. Keep at 37°C before use.
10. 1% Tergazyme (Alconox, White Plains, NY).

2.2.3. Instrumentation

1. Automated microscope for high-content imaging (ImageXpress Micro; Molecular Devices, Sunnyvale, CA; *see* **Note 7**).
2. Bulk-liquid dispenser with plate stackers.
3. Plate washer with plate stackers.
4. Compound-transfer instrument that is able to deliver a nanoliter (nL) range of compounds, such as the CyBi-Well (CyBio) equipped with a 384-pin head.
5. Tissue culture incubators.

3. Methods

3.1. Calcein-AM Viability/Cytotoxicity Assay

1. Cell culture and dispensing cells onto 384-well plates.
   a. Grow adherent cells to near-confluence and avoid overcrowding. Most types of cells grow at 37°C and 5% CO_2.
   b. Perform a trypsin–EDTA digest to remove cells from the flask surface and into suspension. It is critical to disrupt clumps of cells by pipetting the suspension up and down. A well-dispersed suspension allows accurate cell counting, and reduces chances of uneven distribution of cells and clogging of the fine nozzles of the bulk dispenser.
   c. Seed 1,000–5,000 cells per well in 50 μL of medium using a bulk dispenser. To maintain a sterile environment, the dispenser is best placed in a biosafety cabinet with laminar airflow, or in a benchtop enclosure with reversed and filtered airflow. Plates will be delidded in such a clean environment, stacked for bulk dispensing, and relidded. Tubing and other plumbing components can be sterilized with 70% ethanol. To ensure that cells reach the bottom of the wells, the rate of dispensing needs to be adjusted so that it is fast enough to allow the solution to reach bottom of wells without creating air plugs, but gentle enough not to damage the cells. If air plugs prevent cells from settling, centrifugation of the plates usually removes the air plugs and allows the cells to settle properly onto the bottom of the wells.
   d. Incubate seeded plates overnight in tissue-culture incubators.
2. Pin-transfer 100 nL of compound per well. Depending on the stock concentration and experimental design, the final concentration of the compounds can range from approximately 5 to 30 μM (*see* **Notes 8** and **9**).

3. Incubate the compound-treated cells in tissue-culture incubators for the period of time designed to address specific experimental questions (e.g., 24–72 h).
4. Aspirate the compound-containing medium using a plate washer and replace with 50 μL per well of PBS using a bulk dispenser (*see* **Note 10**). Care is needed after cells are incubated with compounds, because cells in certain wells may become fragile due to compound treatment. Aspiration speed and plate-washer pin height (distance between the tips of the pins and the bottom of the wells) both need to be adjusted so that minimal residual liquid will remain, and so that as little cell loss as possible will occur. Again, the dispensing rate on the dispenser needs to be gentle enough not to dislodge the cells.
5. Aspirate the PBS wash with the plate washer and add 30 μL per well of 1-μM calcein-AM in PBS with the bulk dispenser. To make this 1-μM calcein-AM solution, the stock solution of calcein-AM (1 mg/mL, or 1 mM, in DMSO) is diluted 1:1,000 into PBS. Incubate plates for 1 h at room temperature to load calcein-AM into cells (*see* **Note 11**).
6. Wash cells three times with 50 μL/well of PBS, using plate washer to aspirate and bulk dispenser to dispense (*see* **Note 12**).
7. Measure fluorescence intensity on a fluorescence plate reader. Calcein's excitation (ex) and emission (em) maxima are approximately 494/517 nm. However, the green-fluorescein channel (approximately ex/em 485/530 nm) can be used without compromising the signal intensity, if investigators are limited by the predetermined physical properties of the filter sets available to their fluorescence plate readers. When using a filter-based fluorescence plate reader, it is also important to make sure that the proper dichroic mirror is used. A good dichroic mirror eliminates excess excitation light that could interfere with the emission light when no dichroic mirror is used. Also make sure to follow the fluorescence plate reader's steps to optimize plate dimensions, detector height, detector gain, and so on, if such options are offered.

3.2. CFP-YFP-Based FRET Measurement in Conjunction with Fluorescence-Based Nuclear Translocation Monitoring in an Image-Based, High-Content Assay

1. *Cell culture.* Grow sufficient number of C-AR-Y/HEK and HEK293 cells to approach confluence in tissue culture incubators at 37°C and 5% CO2 to be used on "day 1" of the screen. Use trypsin–EDTA to remove the adherent cells and resuspend cells in their respective media.
2. Dispense DHT-containing or DHT-free media onto screening plates. Label the assay plates and a control plate in advance (e.g. using prebarcoded plates). One the screening day, dispense

20 μL of 25-nM DHT in medium to wells on screen plates using a bulk dispenser. Dispense to wells on the control plate as follows: 20 μL respective growth media to columns 1–6 (C-AR-Y/HEK) and columns 13–18 (HEK293) and 20 μL of 25 nM respective growth media containing DHT to columns 7–12 (C-AR-Y/HEK) and columns 19–24 (HEK293).

3. Pin-transfer 100 nL of compounds per well into screen plates. Pin-transfer DMSO into the control plate.
4. Dispense cells into screen plates and the control plate. Using a bulk dispenser, seed C-AR-Y/HEK cells at approximately 10,000 cells per well in 30 μL of medium in screening plates to which compounds have been added. Seed HEK293 and C-AR-Y cells at approximately 10,000 cells per well in 30 μL of respective media into the control plate as follows: C-AR-Y/HEK in columns 1–12 and HEK293 in columns 13–24. Incubate all plates for 24 h at 37°C and 5% CO_2 (*see* **Note 13**).
5. Preparing cells for imaging.
   a. Aspirate the media from the wells in the screening plates using a plate washer.
   b. Fix cells by adding 50 μL of fix solution to each well, and incubate at room temperature for 30 min.
   c. Aspirate the fix solution and wash twice with 50 μL of PBS. Seal the plates with opaque plate seals.
6. Acquire images using the ImageXpress Micro. Four fluorescence channels are needed for a complete set of images. The specific filter sets used for these channels are as follows (excitation/lower transmitted limit of dichroic mirror/emission wavelengths, all in nm): Hoechst (387/415/447), CFP (438/467/483), YFP (500/528/542), and FRET (438/467/542). The Hoechst channel identifies the nucleus, while the other three channels identify the C-AR-Y fusion protein. Both laser- and image-based focusing mechanisms are used to ensure the best focus for each image. The camera is set to 2 × 2 binning of pixels. With a 20× objective, approximately 200 cells can be imaged per field; additional cells can be imaged by acquiring additional fields. Avoid exposure saturation. It takes approximately 45–55 min to image a 384-well plate.
7. Images and data analysis:
   a. *Fluorescence intensity and FRET.* The acquired images are analyzed with MetaExpress (Molecular Devices Corp.). The *Cell Scoring* module is used to generate fluorescence intensity values for CFP, YFP, and FRET channels in both nuclear and cytoplasmic areas in a given well. These per-well values are averaged from all cells sampled in these wells with units of fluorescence units/μm^2.

The HEK293 cells on the control plates provide information regarding background fluorescence intensity for DHT-treated or untreated cells; such background values need to be subtracted from the raw fluorescence reading of C-AR-Y/HEK cells. FRET is usually expressed as a ratio of the signal of the FRET channel over the donor (CFP) channel (*see* **Note 14**).

b. *Nuclear translocation*. The YFP channel is used to track C-AR-Y fusion protein, and the *Translocation* module is used to generate information of percent of cells with nuclear C-AR-Y fusion protein. For the specific C-AR-Y/HEK cell line used, approximately 70–90% of cells undergo nuclear translocation in the presence of DHT, as opposed to under 10% of cells in the absence of DHT.

4. Notes

1. As a cytotoxicity indicator, calcein-AM-based assays provide only live/dead information. They do not suggest specific mechanisms by which cells die or by which cell membranes are compromised.
2. Clear-bottomed plates are ideal to examine cells visually under a microscope. If there is no intention of examining the cells during the assay, opaque-bottomed black plates can also be used. Many vendors' plates are good for this application, such as Nunc and Corning brands.
3. Until recently, CFP and YFP have been the classic and most popular FRET partners for protein applications. In the last few years, there have been reports of alternative FRET partners using DsRed in combination with CFP and GFP, respectively, that demonstrate certain advantages over the CFP-YFP pair *(5, 6)*.
4. AR belongs to the nuclear receptor superfamily and mediates androgenic functions. Upon binding to its ligands (e.g., DHT) in the cytoplasm, AR undergoes a conformational change that allows its N-terminal domain to bind to its C-terminal domain *(7–9)*, followed by homodimerization, entry into the nucleus, and transcriptional activation of target genes *(9, 10)*. Significantly increased FRET signal is generated in the presence of DHT, as opposed to in the absence of DHT, in cells expressing C-AR-Y *(9)*.
5. The HEK293 and C-AR-Y/HEK lines used by the author were kindly provided by Dr. M.I. Diamond, University of California,

San Francisco. The HEK293 line is also available from American Type Culture Collection (ATCC, Manassas, VA).

6. Some brands of plates are better than others for image-based assays, presumably due to the variation in the flatness and thickness of the plastic plate bottom. Investigators are encouraged to test out a few brands with their instrumentation, and to select those that perform the best in their own hands.
7. This protocol should accommodate automated microscopes from other vendors, although different brands have different acquisition settings and use different image analysis software.
8. To reduce protein binding by compounds, growth medium with a high FBS percentage can be changed to one with a lower FBS percentage, such as 0.5% FBS, during compound exposure. This modification will introduce an extra medium-change step, and cells need to be tested to ensure that they will survive such low FBS conditions for the intended duration of compound exposure.
9. Compounds from stock plates are typically in 100% DMSO. During the assay development stage, cells need to be tested to determine if they can tolerate the final DMSO concentration. If cells are found to be sensitive to the intended final DMSO concentration, the volume of cell-culture medium into which the compounds are pinned can be adjusted to reduce the final DMSO concentration to some degree, but this alteration needs to be balanced against the subsequent decrease in compound screening concentrations.
10. It is important to make sure that the plate washer aspirates all wells uniformly in a plate. Occasionally, a few wells in a plate may be missed due to suboptimal vacuum pressure or clogging of individual nozzles. When clogging happens, run maintenance protocols with water, 70% ethanol, and/or 1% Tergazyme to remove clogs.
11. Some cell types may not stand the 1-h incubation in PBS. If cells become unhealthy, membrane integrity may be compromised, and calcein will be released into the solution and lost during the following wash step. Therefore, it is critical to test the loading time for specific cell lines and, if required, to adjust the calcein concentration and incubation duration to achieve optimal results. Alternatively, phenol red-free medium can be used instead of PBS, which in some cell types is better tolerated (T. Gilbert, personal communication).
12. The wash steps promote the reduction of background calcein fluorescence possibly generated by spontaneous hydrolysis of calcein-AM, hydrolysis by residual esterases from serum-supplemented culture medium, or esterases released by a small fraction of cells that are damaged during the calcein-AM loading.

13. The described order of addition for DHT and compounds ensures that the cells will be exposed to both simultaneously. However, investigators can also explore an alternative sequence of addition, i.e., pinning compounds into wells of seeded cells, followed by addition of DHT. This alternative sequence requires one additional day for cell seeding per operation cycle, and, without good scheduling software or planning, this alternative sequence may result in variable times for cell preincubation and compound treatment before ligand stimulation. However, the alternative sequence may help reduce (but not eliminate, due to the relatively long incubation time) the chances of interference with cell attachment by certain compounds. Investigators are advised to balance these pros and cons in selecting the optimal sequence of events based on the properties of cell lines and compound libraries to be used in their specific screens.
14. Sometimes the YFP moiety of a CFP-YFP fusion will generate signal in the FRET channel due to the overlap of the excitation spectra of CFP and YFP, and the physical bandwidth of the excitation/emission filters and the dichroic mirror. This phenomenon is referred to as bleed-through and is instrument-dependent. The amount of bleed-through needs to be subtracted from the total FRET signal. The degree of bleed-through can be determined experimentally for an instrument when a pure YFP sample is measured for its intensities via the instrument's YFP and FRET channels. The fraction of bleed-through is the ratio of signal from the FRET channel over that from the YFP channel.

Acknowledgments

I would like to thank Dr. Bridget Wagner and Ms. Tamara Gilbert for discussions of the calcein-AM protocol, and Drs. Marc I. Diamond and Jeremy Jones for kindly providing the C-AR-Y/HEK and HEK293 cell lines and procedures of a plate reader-based FRET assay for revision and adaptation to the image-based, high-content protocol described here. The work has been funded in whole or in part with Federal funds from the National Cancer Institute's Initiative for Chemical Genetics, National Institutes of Health, under Contract No. N01-CO-12400. The content of this publication does not necessarily reflect the views or policies of the Department of Health and Human Service, nor does mention of trade names, commercial products, or organizations imply endorsement by the US Government.

References

1. Lakowicz, J. R. (ed.) (2006) Principles of Fluorescence Spectroscopy. Springer, New York.
2. Gore, M. (ed.) (2000) Spectrophotometry and Spectrofluorimetry: A Practical Approach. Oxford University Press, Oxford, UK.
3. Valeur, B. (ed.) (2001) *Molecular Fluorescence: Principles and Applications.* Wiley-VCH, Weinheim, Germany.
4. Spence, M. T. Z. (ed.) (2005) *A Guide to Fluorescent Probes and Labeling Technologies.* Invitrogen Corp, Carlsbad, CA.
5. Erickson, M. G., Moon, D. L., and Yue, D. T. (2003) DsRed as a potential FRET partner with CFP and GFP. *Biophys. J.* 85, 599–611.
6. Yang, X., Xu, P., and Xu, T. (2005) A new pair for inter- and intra-molecular FRET measurement. *Biochem. Biophys. Res. Commun.* 330, 914–920.
7. He, B., Kemppainen, J. A., Voegel, J. J., Gronemeyer, H., and Wilson, E. M. (1999) Activation function 2 in the human androgen receptor ligand binding domain mediates interdomain communication with the NH2-terminal domain. *J. Biol. Chem.* 274, 37219–37225.
8. Wong, C. I., Zhou, Z. X., Sar, M., and Wilson, E. M. (1993) Steroid requirement for androgen receptor dimerization and DNA binding. *J. Biol. Chem.* 268, 19004–19012.
9. Schaufele, F., Carbonell, X., Guerbadot, M., Borngraeber, S., Chapman, M. S., Ma, A. A. K., Miner, J. N., and Diamond, M. I. (2005) The structural basis of androgen receptor activation: Intramolecular and intermolecular amino-carboxy interactions. *Proc. Natl. Acad. Sci. USA* 102, 9802–9807.
10. Lee, H. J. and Chang, C. (2003) Recent advances in androgen receptor action. *Cell. Mol. Life Sci.* 60, 1613–1622.

Chapter 8

Reporter Gene Assays

Andy M.F. Liu, David C. New, Rico K.H. Lo, and Yung H. Wong

Summary

Reporter gene assays are versatile and sensitive methods of assaying numerous targets in high-throughput drug-screening programs. A variety of reporter genes allow users a choice of signal that can be tailored to the required sensitivity, the available detection apparatus, the cellular system employed, and the required compatibility with multiplexed assays. Promoters used to drive reporter gene expression can be activated either by a broad range of biochemical pathways or by the selective activation of individual targets. In this chapter, we will introduce some of the considerations behind the choice of reporter gene assays and describe the methods that we have used to establish 96-well format luciferase and aequorin assays for the screening of ligands for G protein-coupled receptors.

Key words: Aequorin, Chimera, CRE, GPCR, G-protein, Luciferase, NF-kappa-B, Receptor, Reporter gene, STAT.

1. Introduction

The development of cell-based high-throughput screening platforms owes its success to the ability to manipulate the genetic and biochemical composition of eukaryotic cells. Notably, cells can be engineered to express specific gene products in response to a given stimulus. The gene product itself may possess an inherent property that enables it to be measured directly, e.g., green fluorescent protein (GFP), or it may display enzymatic activity that can be monitored, e.g., luciferase. Alternatively, the gene product may respond to changes in the levels of a signaling molecule, e.g., the Ca^{2+} ion-mediated activation of aequorin luminescence. The continuing development of genetically engineered vectors, coupled with powerful detection techniques, allows many cell types

P.A. Clemons et al. (eds.), *Cell-Based Assays for High-Throughput Screening, Methods in Molecular Biology, Vol. 486*

DOI: 10.1007/978-1-60327-545-3_1

to be engineered as biodetectors for myriad classes of biochemical and signaling pathways.

The reporter gene unit consists of a promoter and the reporter gene. When establishing an assay system, a number of factors need to be considered in order to optimize the assay. The choice of promoter, the number of copies of the promoter, and the nature of the reporter gene will allow control of the basal level of reporter gene activity, control of the specificity of activation, and control of the degree of stimulation measured *(1)*. Endogenous promoters, such as c-fos, the cAMP response element (CRE), or the estrogen response element, are commonly used, but may suffer interference from endogenous intracellular signaling events. However, their regulation by multiple signaling pathways makes them widely applicable, even allowing drugs that activate different signaling events to be identified in the same assay format. Alternatively, exogenous promoters, such as the yeast Gal4 response element system, can be used to reduce unwanted activation by native transcription factors *(2)*. There are many examples of the optimization and employment of a variety of promoters in high-throughput screening programs *(3)*.

The reporter gene itself should ultimately generate a signal that can be clearly identified. The reporter gene products can be either intracellular or extracellular in nature. Intracellular products are retained in the cell for quantification in situ or following cell lysis. Extracellular products are secreted into the extracellular medium for assay, allowing repeated experimentation and sampling without disturbing the cells. Commonly used intracellular reporter genes are chloramphenicol acetyltransferase (CAT), β-galactosidase, luciferase, aequorin, and GFP. Extracellular reporter genes are usually secreted placental alkaline phosphatase (SPAP) or β-lactamase *[(3)* and references therein]. Each reporter gene has its advantages and disadvantages in terms of sensitivity, time taken for signal generation, ease of signal detection (e.g., products can be radioactive, fluorescent, bioluminescent, etc.), cost, miniaturization, and automation. A further consideration may also be the compatibility within dual reporter gene assays or multiplexed assays.

A recent report of a dual reporter gene assay demonstrated the ease with which signals from two different classes of G protein-coupled receptors (GPCRs) can be measured simultaneously. Two cell lines were generated that expressed either CRE coupled to firefly luciferase (CRE-luc) or to a 12-*O*-tetradecanoyl-phorbol -13-acetate (TPA)-response element coupled to *Renilla* luciferase. Coseeding of the reporter cell lines allowed independent or simultaneous measurement of ligand activation of GPCRs that activate distinct intracellular signaling pathways *(4)*. The authors have also suggested that the inclusion of a third GFP-based reporter construct would allow three separate GPCRs to be coassayed. Reporter gene assays have also proven amenable to inclusion in multiplexed assays that also measure Ca^{2+} mobilization in a 384-well format. For example, by using the β-lactamase-based GeneBLAzer® reporter gene system (Invitrogen, Carlsbad, CA) and loading the same cells with the calcium-sensing Fluo-4 NW dye, researchers were able to rapidly verify positive *hits* in a drug screen by correlating the data from Ca^{2+}-based assays and reporter gene assays performed sequentially on the same cells *(5)*. Running these two assays also facilitated the rapid elimination of false positives generated by either assay format, allowing huge savings in cost and time to be made.

In our experience, reporter gene-based assays are an effective way for laboratories with limited resources to develop and deploy innovative and sensitive high-throughput screening programs. Reporter gene assays can be applied to a huge range of targets including GPCRs, nuclear receptors, receptor tyrosine kinases, enzymes, and transcription factors. Furthermore, by tailoring the incorporation of relevant promoters and reporter genes into different assays, diverse screening programs can be established that utilize common detection equipment, increasing the cost efficiency of drug screening.

We use both luciferase and aequorin reporter genes. The luciferase genes are driven by STAT3, NFκB, or CRE promoters. These allow us to assay most GPCRs, with G_s- and $G_{i/o}$-coupled GPCRs effectively regulating CRE-driven luciferase activity by modulating intracellular cAMP levels *(6)*. G_q- and $G_{i/o}$-coupled GPCRs are assayed by either STAT3- or NFκB-driven assays, while $G_{12/13}$-coupled receptors generally promote NFκB-driven luciferase activity *(7–9)*. The flexibility of all of our reporter gene assays is greatly enhanced by the incorporation of promiscuous or chimeric G protein α-subunits, notably $G\alpha_{16}$, 16z25, or 16z44 *(10)*. These subunits allow virtually all GPCRs tested to date to promote the activation of phospholipase Cβ (PLCβ)/inositol trisphosphate (IP_3)/Ca^{2+} pathways, enabling a single assay to be used to screen for GPCR activity. Numerous chimeras have been stably or transiently expressed in a variety of cell lines and incorporated into a number of other assay platforms, including FLIPR, microarrayed compound screening, and yeast autocrine selection *(11, 12)*. In fact, cell lines stably expressing GPCRs and promiscuous G proteins are commercially available (Chemicon, Temecula, CA). Using endogenous and chimeric G proteins, we and others have successfully incorporated dozens of GPCRs into aequorin-based reporter gene screening programs *(13–15)*. In this chapter, we will detail the methods that we regularly use in our own GPCR analysis and screening programs.

2. Materials

1. Dulbecco's Modified Eagle Medium (DMEM) (Gibco/Invitrogen/Invitrogen, Carlsbad, CA).
2. Modified Eagle Medium (MEM) (Gibco/Invitrogen).
3. OPTI-Modified Eagle Medium (Opti-MEM, Gibco/Invitrogen).
4. Phenol red-free DMEM (Sigma-Aldrich; St. Louis, MO).
5. Fetal bovine serum (FBS; Gibco/Invitrogen).
6. Dulbecco's phosphate-buffered saline (PBS; Gibco/Invitrogen).
7. Hanks' balanced salt solution (HBSS), with or without calcium chloride (Gibco/Invitrogen).
8. Solution of trypsin (1×) is diluted from trypsin (10×) with PBS (Gibco/Invitrogen).
9. Solution of penicillin/streptomycin (20×) (Gibco/Invitrogen).
10. Zeocin (250mg/mL) (Gibco/Invitrogen).
11. Geneticin (G418; Calbiochem, San Diego, CA) is dissolved in Opti-MEM at 100mg/mL and stored at –20°C. The working concentration is 50mg/mL.
12. Lipofectamine PLUS and Lipofectamine 2000 reagents (Gibco/Invitrogen).
13. 2× HEPES-buffered saline (HBS): 280mM NaCl, 10mM KCl, 1.5mM Na_2HPO_4, 12mM dextrose, and 50mM HEPES (pH 7.05). The solution should be sterilized by passing through a 0.2-µm filter.
14. Multichannel pipettor, Multipette Plus dispenser and Combitips Plus (5mL and 2.5mL) (Eppendorf, Westbury, NY).
15. Tissue-treated 96-well polystyrene white plates with flat bottoms (Nalge Nunc, Rochester, NY).
16. Polypropylene 96-well round-bottom block and solvent reservoir (USA Scientific, Ocala, FL).
17. Conical 96-well polypropylene plates (Corning Costar, Lowell, MA).
18. Mitochondrial apoaequorin expression vector and coelenterzine *f* (Invitrogen).
19. Luciferase reporter gene constructs with STAT3-, CREB-, and NFκB-response elements (pSTAT3-Luc, pCRE-Luc, and pNFkB-Luc; Clontech, Mountain View, CA).

20. Luciferin and luciferase lysis buffer are from a luciferase assay kit (Roche, Mannheim, Germany).
21. 96-well format luminometer (EG&G Berthold, Bundoora, Australia).

3. Methods

3.1 Luciferase Assay

3.1.1. Preparation of Cells Stably Expressing CRE- or NFκB-Driven Luciferase Reporter Genes

To minimize the variations in different sets of experiments run on different days, stable cell lines carrying CRE- or NFκB-driven luciferase reporter genes have been generated. The plasmids carrying these reporter genes do not encode any selection markers. Therefore, we cotransfect pcDNA3 to provide a means to select cells carrying both the reporter gene constructs and pcDNA3. We have found that it is not necessary to isolate a clonal cell line. Cells that survive the selection process are harvested and provide robust, consistent responses almost indefinitely. After the establishment of the stable cell lines, cDNAs encoding GPCRs and G proteins can be transiently introduced into the cells prior to the assays (**Subheading 3.1.2**).

1. Seed 100-mm plates of 2×10^6 HEK293 cells in MEM medium containing 10% FBS and 1× penicillin/streptomycin (*see* **Note 1**).
2. The following day, the medium should be exchanged for 9-mL fresh MEM/10% FBS/1× penicillin/streptomycin.
3. After 3h, prepare calcium phosphate transfection buffers as described here (the indicated volumes are required for each plate of cells):
   (a) Aliquot 500μL 2× HBS into a 5-mL sterile plastic tube.
   (b) Prepare the DNA mix in a second tube containing 50μL of 2.5M $CaCl_2$ (sterilized through a 0.2-μm filter), 9μg of reporter gene cDNA, and 1μg of selection marker DNA (pcDNA3), and make up volume to 500μL with H_2O.
   (c) Add DNA mix slowly to 2× HBS by pipetting dropwise.
   (d) Use a pipette to bubble air though the solution for (30 seconds) to allow complete mixing.
   (e) Incubate at room temperature for 20min.
   (f) Add the mix to the cells dropwise with gentle swirling.
4. Incubate for 16h, 37°C, 5% CO_2.
5. Aspirate the transfection medium, wash transfected cells with PBS three times to completely remove calcium phosphate-DNA

precipitates, and replenish with fresh MEM/10% FBS/1× penicillin/streptomycin.

6. Incubate at 37°C, 5% CO_2 overnight.
7. Aliquot 50μL of 100mg/mL G418 to each 100-mm plate for selection, for a final concentration of 500μg/mL G418.
8. Fresh medium with 50μL of 100mg/mL G418 (to give a concentration of 500μg/mL) is replenished every 3–4 days, after washing three times with PBS, until selection is finished (*see* **Note 2**).
9. Cells stably carrying luciferase reporter are maintained in 250μg/mL G418.

*3.1.2. Transient Transfection of GPCR and G Protein cDNAs into Cell Lines Stably Carrying Luciferase Reporter Gene Constructs (see **Note 3**)*

1. 15,000 HEK293 cells carrying luciferase reporter genes (*see* **Subheading 3.1.1**) are seeded per well of a 96-well white microplate using MEM (10% FBS).
2. The following day prepare the transfection cocktail as described here:
   (a) Mix 12.5-ng GPCR cDNA and 37.5-ng G protein cDNA per well.
   (b) Dilute 0.2μL of PLUS reagent with 25-μL Opti-MEM per well.
   (c) Add PLUS/Opti-MEM mixture to the cDNA.
   (d) Incubate at room temperature for 15min.
   (e) Dilute 0.2-μL Lipofectamine reagent with 25-μL Opti-MEM per well.
   (f) Add the diluted Lipofectamine to the cDNA/PLUS/Opti-MEM solution.
   (g) Incubate at room temperature for a further 15min.
3. Remove the growth medium from the white microplate seeded with the reporter stable cell line.
4. Aliquot 50μL of the transfection cocktail to each well of the white microplate.
5. Incubate at 37°C for 3h.
6. Add 25μL of Opti-MEM containing 30% FBS.
7. Further incubate at 37°C overnight.
8. Proceed to **Subheading 3.1.4** for assay.

*3.1.3. Preparation of Cells Transiently Expressing STAT3 Promoter-Driven Luciferase Reporter Gene Constructs (see **Note 3**)*

In contrast to the CRE promoter- and NFκB promoter-driven luciferase reporter genes, attempts to establish HEK293 cell lines stably expressing STAT3 promoter-driven luciferase reporter genes have resulted in luciferase signals that diminish with each passage of the cells. Therefore, transient introduction of the

STAT3-driven luciferase reporter gene, along side the GPCR and G protein cDNAs, is employed.

1. 15,000 HEK293 cells are seeded per well in a 96-well white microplate with MEM (10% FBS).
2. The following day, prepare the transfection cocktail following **step 2** of **Subheading 3.1.2**, with a modification to **step 2a**.
   (a) Mix 10-ng GPCR cDNA, 10-ng G protein cDNA, and 100-ng STAT3 reporter cDNA per well.
3. Remove growth medium from the white microplate seeded with HEK293 cells.
4. Distribute 50-μL Opti-MEM to each well.
5. Aliquot 50μL of the transfection cocktail to each well.
6. Incubate at 37°C for 3h.
7. Add 25μL of Opti-MEM containing 30% FBS.
8. Further incubate at 37°C overnight.
9. Proceed to **Subheading 3.1.4** for assay.

3.1.4. Ligand Induction

Ligand Induction for NFκB- and STAT3-Luciferase Assays (*see* ***Note 4***)

1. Prepare and dilute appropriate agonists and/or antagonists using phenol red-free DMEM (50μL will be required for each well of a 96-well plate).
2. Remove the transfection cocktail from the 96-well white microplate.
3. Aliquot 50μL of the prepared solutions with ligands or phenol red-free DMEM medium (as a basal control) to appropriate wells. For determination of the contribution of $G\alpha_{i/o}$ proteins to a response, wells should be pretreated with 100ng/mL PTX (in phenol red-free DMEM) for 4h before ligand addition. The PTX can be added directly to the transfection cocktail.
4. Incubate at 37°C for 16h.
5. Remove the medium completely.
6. Aliquot 25-μL luciferase lysis buffer per well and store at –80°C until use in the luciferase assay (*see* **Subheading 3.1.5**).

Agonist Induction for CRE Assays (*see* ***Note 4***)

1. Prepare diluted ligands and controls using phenol red-free DMEM.
2. Remove the transfection cocktail from the 96-well white microplate.
3. Aliquot 50μL of the prepared ligand solutions or controls, as appropriate, to the wells. Should forskolin be required, it should also be included at this point at a final concentration of 10μM (*see* **Note 5**).

4. Incubate for:
   (a) To measure activation of adenylyl cyclase by G_s-coupled GPCRs: overnight at 37°C.
   (b) To measure inhibition of adenylyl cyclase activity by $G_{i/o}$-coupled receptors: 30min at 37°C. Remove stimulus, including forskolin, completely and wash twice with 100-μL phenol red-free DMEM. Replenish with 50-μL phenol red-free MEM and incubate at 37°C for 16h.
5. Remove medium completely.
6. Aliquot 25-μL luciferase lysis buffer per well and store at –80°C until use in the luciferase assays (*see* **Subheading 3.1.5**).

3.1.5. Luciferase Assay (see ***Note 6****)*

1. Prepare luciferin solution and 1× luciferase lysis buffer, and equilibrate both to room temperature (25μL of each solution will be required per well).
2. Prewarm the luminometer to 25°C. Ensure that the 96-well plates have been removed from –80°C and thawed.
3. Charge injector P, of the 96-well luminometer (EG&G Berthold, Bundoora, Australia), with luciferin solution and injector M with luciferase lysis buffer.
4. Inject 25-μL luciferase lysis buffer by injector M followed by 25-μL luciferin solution from Injector P to each well of the luminescent white plate.
5. Monitor luminescence released for 10s after luciferin addition. Examples of the data generated by these luciferase reporter assays can be seen in **Figs. 1** and **2**.

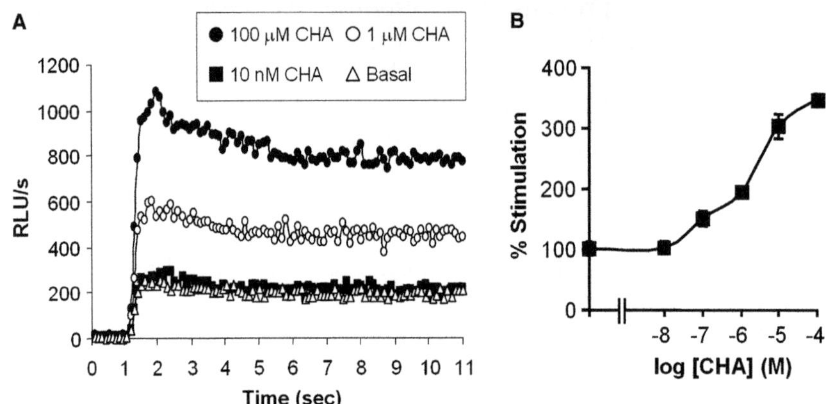

Fig. 1. Dose-response curve of N^6-cyclohexyladenosine (CHA, selective agonist for adenosine A_1 receptor) for $A_1R/G\alpha_{16}$-transfected HEK293 cells stably carrying a NFκB-driven luciferase reporter construct. To eliminate $G_{i/o}$ signal transduction, transfectants were pretreated with pertussis toxin (100 ng/mL) for 4 h prior to agonist induction. Different concentrations of CHA were added to the cells for 16 h, and the luciferase activities of the cell lysates were measured. (**A**) Triplicates were performed and the results from one set of wells are illustrated as representative. (**B**) Construction of the dose-response curve was accomplished by plotting the % stimulation corrected to the basal vs. the logarithmic concentrations of CHA. Data shown represent the mean ± SE performed in triplicate.

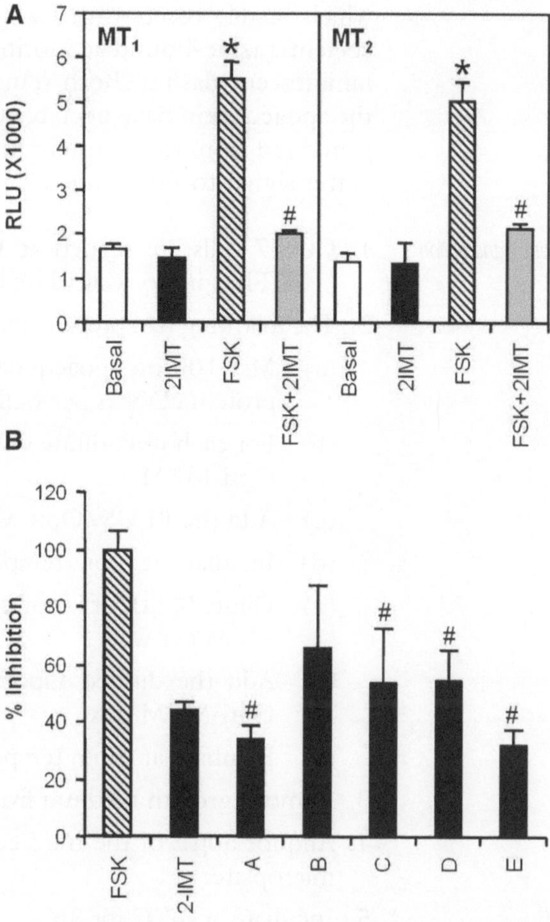

Fig. 2. The inhibitory responses of melatonin type 1 and type 2 receptors (MT_1 and MT_2) coupled to $G_{i/o}$ proteins as measured by a CRE-driven reporter gene. HEK293 cells stably carrying a CRE-driven luciferase reporter construct were transfected with MT_1 or MT_2 and the transfectants were challenged with plain medium (basal), 2-iodomelatonin (an agonist for MT_1 and MT_2; 2-IMT, 10 μM), forskolin (FSK, 10 μM), or FSK with 2-IMT. (**A**) Raw integrated RLUs were plotted and data shown represent the mean ± SE performed in triplicate. (**B**) A panel of 2-IMT derivatives (A–E) was exposed to MT_1-transfected HEK293 cells stably carrying the CRE-driven luciferase reporter. The magnitudes of % inhibition of the forskolin response (where FSK stimulation alone was set to 100%) were plotted and data shown represent the mean ± SE performed in triplicate. *, agonist-induced RLU is significantly higher than the basal level (Dunnett's *t* test, $p < 0.05$); #, agonist-induced response was significantly inhibited as compared to FSK (Dunnett's *t* test, $p < 0.05$).

3.2. Aequorin Screening

This screening is primarily made possible by the coupling of GPCRs to G proteins, leading to downstream PLCβ activation, IP_3 accumulation, and Ca^{2+} release from endoplasmic reticulum. The released Ca^{2+} is quantified by the presence of Ca^{2+}-sensitive aequorin photoprotein. The aequorin complex is composed of apoaequorin, coelenterazine (a cofactor), and oxygen, which readily reacts with Ca^{2+}, causing the degradation of the coelenterazine-bound aequorin, generating coelenteramide and luminescent flashes. Both transient and stable transfections of the apoaequorin have been tested, but we find that the response generated from the transient introduction of apoaequorin gives better signal-to-noise ratios.

3.2.1. Transient Transfection (see Note 3)

1. COS-7 cells are seeded at 10,000 cells/well using DMEM (10% FBS) in 96-well white luminescent plates.
2. The following day, prepare transfection cocktail as described here:
   (a) Mix 100-ng apoaequorin cDNA, 50-ng GPCR, and G protein cDNAs per well.
   (b) For each well, dilute 0.2μL of PLUS reagent with 25-μL Opti-MEM.
   (c) Add the PLUS/Opti-MEM diluent to the cDNA mix.
   (d) Incubate at room temperature for 15min.
   (e) Dilute 0.2μL of Lipofectamine reagent with 25-μL Opti-MEM per well.
   (f) Add the diluted Lipofectamine to the cDNA/PLUS/Opti-MEM mix.
   (g) Incubate at room temperature for a further 15min.
3. Remove growth medium from the 96-well microplate.
4. Aliquot 50μL of the transfection cocktail to each well of the microplate.
5. Incubate at 37°C for 3h.
6. Add 50-μL Opti-MEM containing 4% FBS (*see* **Note 7**).
7. Incubate at 37°C for 48h.

3.2.2. Apoaequorin Labeling (see Note 8)

1. For each well prepare labeling medium consisting of 100-μL calcium-free HBSS (pH 7.5) with 2.5μM coelenterazine *f* and 20mM HEPES (pH 7.4).
2. The transfection medium is removed from the 96-well microplate and replaced by 100μL of the labeling medium.
3. Place the microplate in a dark, humidified incubator at 37°C for 4h.

3.2.3. Aequorin Assay (see Note 9)

1. The luminometer should be prewarmed to 37°C.
2. Prepare appropriate controls and drug solutions in 3× concentrations (50μL per well will be required).
3. Warm the drug solutions to 37°C.
4. Set Injector P to inject 50-μL control or drug solutions.
5. Following injection, monitor the luminescence released for 15s. Examples of the types of data generated by aequorin assays can be seen in **Figs. 3** and **4**.

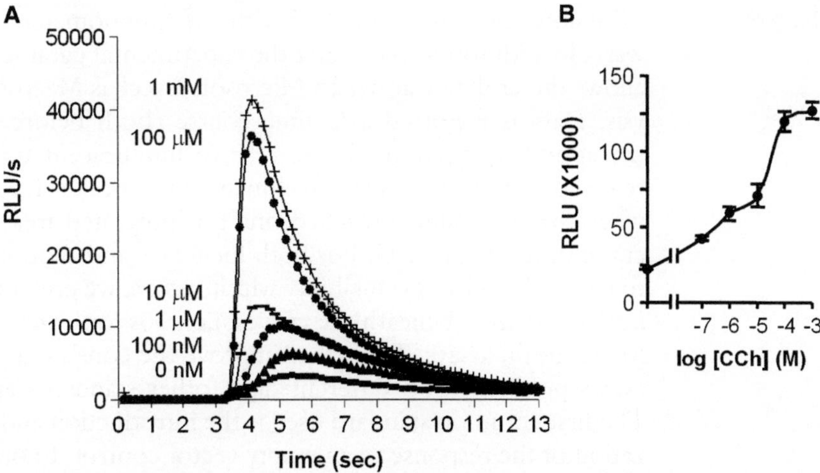

Fig. 3. Dose-response curves of carbachol for COS-7 cells transiently transfected with muscarinic type 1 receptor (M_1R) and apoaequorin. Transfectants were labeled with 2.5 μM coelenterazine *f* for 4 h. (**A**) Serial dilutions of carbachol were prepared and screening was performed in increasing order of carbachol concentrations. Triplicates were performed and the results from one set of wells are illustrated as representative. (**B**) Construction of a dose-response curve was accomplished by plotting the total integrated RLUs vs. the logarithmic concentrations of carbachol. Data shown represent the mean ± SE performed in triplicate.

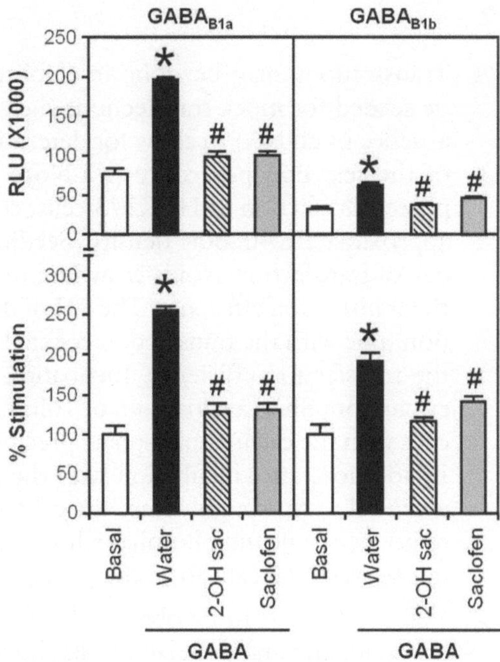

Fig. 4. Aequorin assay of the effects of $GABA_B$-selective antagonists in COS-7 cells transfected with $GABA_{B2}$ and either $GABA_{B1a}$ or $GABA_{B1b}$ receptor subunits, as well as the G protein α-subunit 16z25. Transfectants were treated with basal (*open bar*), GABA (500 μM; *closed bar*), GABA with 2-hydroxy saclofen (2-OH Sac; 50 μM; *diagonally striped bar*), and GABA with saclofen (50 μM, *gray bar*). 2-OH Sac and saclofen are antagonists at $GABA_B$ receptors and they were added simultaneously with GABA. Raw integrated RLUs and % stimulation (as compared to their corresponding basals) were plotted. Data shown represent the mean ± SE performed in triplicate. *Agonist-induced RLU is significantly higher than the basal level (Dunnett's t test, $p < 0.05$); #, antagonist significantly suppressed the GABA-induced Ca^{2+} mobilization (Dunnett's t test, $p < 0.05$).

3.3. Data Analysis

WinGlow software is supplied with the luminometer. This software, in addition to recording the experimental parameters used, allows direct data transfer to Microsoft Excel as Macros for analysis. Data is exported as a time course (both before and after the stimulus injection) of the relative luminescent unit (RLU) released. The readouts in Excel include the peak value, the time when the peak value is reached, and the integrated area under the curve in terms of RLU. For both luciferase and aequorin assays, to ensure better reproducibility within assays, we prefer to rely on integrated areas beneath the curves, i.e., 10s for luciferase or 15s for aequorin assays. To further improve the consistency between assays performed on different days, other standards are useful. The first common standard used is the introduction and normalization of the response to an empty vector control. In other cases, for instance, when the efficacy of an agonist is to be evaluated, the fold-stimulation or percent-stimulation over a blank control is used.

4. Notes

1. Transfections may be done in duplicate. Plates should also be seeded for mock transfections (i.e., cells transfected in the absence of cDNA) to allow for determination of the endpoint of the selection procedure (*see* **Note 2**). For calcium phosphate transfection of HEK293 cells, cells should be seeded at approximately 40–60% density. Seeding cells just before the day of transfection avoids crowding of cells, which may lower the transfection efficiency. The pH of the HBS and the incubation time with the transfection cocktail play significant roles in the transfection efficiency. Incubation time periods normally range from 8h to a maximum of 16h. Prolonged incubation of cells with the calcium phosphate precipitate induces cell death. In addition, after incubation with the transfection cocktail, it is crucial to thoroughly rinse the cells with PBS to completely remove the calcium phosphate; however, care should be taken not to wash the cells from the plate.
2. Cell death starts to be observable after G418 has been introduced for the first 4–5 days. Cells should be washed with PBS and new G418 should be replenished into the fresh medium every few days. Normally, the whole selection process takes 2–3 weeks to finish, at which point all cells in the mock transfected plates should be dead and any living cells in the transfected plates can be assumed to be stably expressing the reporter gene constructs.

3. Triplicates are always performed in our 96-well format assays, and often many 96-well plates are transfected with the same cDNAs. Hence, instead of mixing cDNAs for individual wells, *master mixes* for a number of wells are prepared to reduce the number of pipetting steps and possible pipetting errors. A multichannel pipettor is employed to ease the process of removing growth medium and adding transfection cocktail to microplates. Complete removal of the growth medium before addition of the transfection cocktail is essential, as serum in the medium hinders the transfection by LipofectaminePLUS reagents.
4. As phenol red has been shown to have an effect on the luciferin reaction, phenol red-free medium is used for drug treatment after transfection. In addition, the transfection cocktail must be thoroughly removed to avoid activation of the response elements by growth factors and hormones, among other things, in the serum. At least one freeze–thaw cycle is required for complete cell lysis by the luciferase lysis buffer. Thus, the collected cell lysates can be kept at −80°C until required for the luciferase assays.
5. For the CRE-driven luciferase reporter, two different experimental designs are optimized for the evaluation of either the activation or inhibition of adenylyl cyclase through the regulation of G_s or $G_{i/o}$ proteins, respectively. For the activation of G_s proteins, the experimental setup is similar to that for STAT3- or NFκB-driven luciferase assays. However, for measurements of $G_{i/o}$ protein-mediated activity, the activation of $G_{i/o}$ proteins is revealed through their inhibition of adenylyl cyclase activity. Adenylyl cyclase can be activated either directly by forskolin or through the activation of G_s proteins by a G_s-coupled GPCR. In most of our studies, forskolin is used, as it guarantees robust stimulation of adenylyl cyclase. This necessitates the use of extra controls. Generally, we use basal agonist for the receptor under investigation, forskolin alone, and forskolin with the receptor's agonist. The washing step is particularly important in CRE-driven luciferase assays using forskolin. Traces of forskolin induce remarkable CRE-driven luciferase activity and overwhelm the inhibitory effects mediated by the $G_{i/o}$ proteins. Complete removal of residual forskolin is accomplished by careful aspiration of stimuli and additional washing steps with plain medium. Alternatively, the activation of a G_s-mediated pathway can be achieved by cotransfection of a G_s-coupled receptor. This G_s-mediated activation of adenylyl cyclase is generally weaker than the forskolin-mediated stimulation, and may be preferred in the examination of G_i-coupled receptors that couple weakly to adenylyl cyclase.

6. To maintain consistency between runs, the luciferin substrate solution, the lysis buffer, and the microplate reader are always prewarmed to ambient temperature before assays. An additional 25-μL lysis buffer is injected to ensure that the samples of cell lysates are equilibrated to ambient temperature and the cell lysate is evenly mixed and distributed throughout the well.
7. The growth rate for COS-7 cells is faster than for HEK293 cells; thus, a final concentration of 2% fetal bovine serum is used for the replenishment after transfection.
8. Activated aequorin, i.e., coelenterazin *f*-bound apoaequorin, reacts quickly with Ca^{2+} for luminescence release. Therefore, it is absolutely essential to maintain a Ca^{2+}-free environment starting from the labeling step.
9. Since intact cells are used for the aequorin assays, the residing platform inside the luminometer, the microplate, and the plastic tubes with the drug solutions are warmed to 37°C. A sudden temperature change for charged cells could cause stress and induce inaccurate readings. Since only one injector (Injector P) is available and suitable in MicroLumat (Berthold) for drug distribution, the screening itself always starts with wells for the basal measurements followed by wells with agonists. Otherwise, wash steps are necessary to completely flush out ligands left in the injector tubing.

Acknowledgments

This work was supported in part by the Hong Kong Jockey Club and grants from the Research Grants Council of Hong Kong (HKUST 6420/05M) and the University Grants Committee (AoE/B-15/01). YHW was a recipient of the Croucher Senior Research Fellowship.

References

1. Ponglikitmongkol, M., White, J., and Chambon, P. (1990) Synergistic activation of transcription by the human estrogen receptor bound to tandem responsive elements. *EMBO J.* 9, 2221–2231.
2. Miller-Martini, D. M., Chan, R. Y. K., Ip, N. Y., Sheu, S. J., and Wong, Y. H. (2001) A reporter gene assay for the detection of phytoestrogens in traditional Chinese medicine. *Phytotherapy Res.* 15, 487–492.
3. New, D. C., Miller-Martini, D. M., and Wong, Y.H. (2003) Reporter gene assays and their applications to bioassays of natural products. *Phytother. Res.* 17, 439–448.
4. Kent, T. C., Thompson, K. S., and Naylor, L. H. (2005) Development of a generic dual-reporter gene assay for screening G-protein-coupled receptors. *J. Biomol. Screen.* 10, 437–446.
5. Hanson, B. J. (2006) Multiplexing Fluo-4 NW and a GeneBLAzer transcriptional assay for high-throughput screening of G-protein-coupled receptors. *J. Biomol. Screen.* 11, 644–651.

6. Do, E. U., Piao, L. Z., Choi, G., Choi, Y. B., Kang, T. M., Shin, J., Chang, Y. J., Nam, H. Y., Kim, H. J., and Kim, S. I. (2006) The high throughput screening of neuropeptide FF2 receptor ligands from Korean herbal plant extracts. *Peptides* 27, 997–1004.
7. Lo, R. K. and Wong, Y. H. (2004) Signal transducer and activator of transcription 3 activation by the δ-opioid receptor via Gα14 involves multiple intermediates. *Mol. Pharmacol.* 65, 1427–1439.
8. Liu, A. M. and Wong, Y. H. (2005) Activation of nuclear factor κB by somatostatin type 2 receptor in pancreatic acinar AR42J cells involves Gα14 and multiple signaling components: A mechanism requiring protein kinase C, calmodulin-dependent kinase II, ERK, and c-Src. *J. Biol. Chem.* 280, 34617–34625.
9. Lin, P. and Ye, R. D. (2003) The lysophospholipid receptor G2A activates a specific combination of G proteins and promotes apoptosis. *J. Biol. Chem.* 278, 14379–14386.
10. Mody, S. M., Joshi, S. A., Ho, M. K. C., and Wong, Y. H. (2000) Incorporation of Gαz-specific sequence at the carboxy terminus increases the promiscuity of Gα16 towards Gi-coupled receptors. *Mol. Pharmacol.* 57, 13–23.
11. New, D. C. and Wong, Y. H. (2004) Characterization of CHO cell lines stably expressing a Gα16/z chimera for high throughput screening of GPCRs. *Assay Drug Dev. Technol.* 2, 269–280.
12. New, D. C. and Wong, Y. H. (2005) Chimeric and promiscuous G proteins in drug discovery and the deorphanization of GPCRs. *Drug Des. Rev. Online.* 2, 66–79.
13. Liu, A. M. F., Ho, M. K. C., Wong, C. S. S., Chan, J. H. P., Pau, A. H. M., and Wong, Y. H. (2003) Gα16/z chimeras efficiently link a wide range of G protein-coupled receptors to calcium mobilization. *J. Biomol. Screen.* 8, 39–49.
14. Ungrin, M., Singh, L., Stocco, R., Sas, D., and Abramovitz, M. (1999) An automated aequorin luminescent-based functional calcium assay for GPCRs. *Anal. Biochem.* 272, 34–42.
15. Le Poul, E., Hisada, S., Mizuguchi, Y., Dupriez, V. J., Burgeon, E., and Detheux, M. (2002) Adaptation of aequorin functional assay to high throughput screening. *J. Biomol. Screen.* 7, 57–65.

Chapter 9

Screening for Chemical Inhibitors of Heterologous Proteins Expressed in Yeast Using a Simple Growth-Restoration Assay

Aruna D. Balgi and Michel Roberge

Summary

Overexpression of heterologous proteins in the yeast *Saccharomyces cerevisiae* often inhibits its growth, while inhibitors of the overexpressed proteins can restore growth. These simple observations form the basis of a technically easy, inexpensive, scalable, and widely applicable assay to identify inhibitors of such proteins. An expression plasmid for the inducible expression of a gene of interest is introduced into a yeast strain rendered more sensitive to chemicals by deletion of efflux pumps. Protein expression is induced, cells are exposed to test chemicals, and growth is measured by A_{600} reading. The chemicals that relieve growth inhibition are subjected to secondary assays to establish their selectivity toward the protein of interest. This assay has been used successfully to identify inhibitors of proteins of viral, microbial, and mammalian origin.

Key words: Chemical inhibitor, Drug screening, Growth restoration, *Saccharomyces cerevisiae*.

1. Introduction

Identifying chemical inhibitors of proteins can be a difficult task, requiring the development of in vitro or cell-based assays tailored to the specific activity of a protein target. The assays may be complex or necessitate expensive reagents, making it difficult to scale them up to the level of throughput needed to screen large chemical libraries. Traditional drug-discovery efforts have therefore tended to focus on enzymes and small-ligand receptors, which are amenable to in vitro high-throughput activity or ligand-binding assays. Many other protein classes playing major roles in disease have been overlooked in drug discovery because

P.A. Clemons et al. (eds.), *Cell-Based Assays for High-Throughput Screening, Methods in Molecular Biology, Vol. 486*

DOI: 10.1007/978-1-60327-545-3_1

it is not obvious how to assay them in high-throughput screens. These include structural proteins, regulatory proteins, effector proteins, proteins with ill-defined functions, or proteins that are difficult to express in vitro.

It has been observed that overexpression of heterologous proteins in the yeast *Saccharomyces cerevisiae* often inhibits growth *(1)*. Inhibitors of these proteins can restore yeast growth, providing a simple generic drug-screening assay. This type of assay has been used to identify an inhibitor of the influenza virus M2 envelope protein, which is an ion channel *(2)*, inhibitors of human poly(ADP)ribose polymerase *(3)*, and inhibitors of human indoleamine dioxygenase *(4)*. Tugendreich et al. have observed growth inhibition for 30% of human proteins overexpressed in yeast *(1)*, an observation corroborated by our own studies. For proteins that do not inhibit growth in standard culture media, it may be possible to alter culture conditions to reveal a growth-inhibitory phenotype *(4)*.

The basic assay, represented diagrammatically in **Fig. 1**, consists of inserting the coding sequence of a gene of interest into a

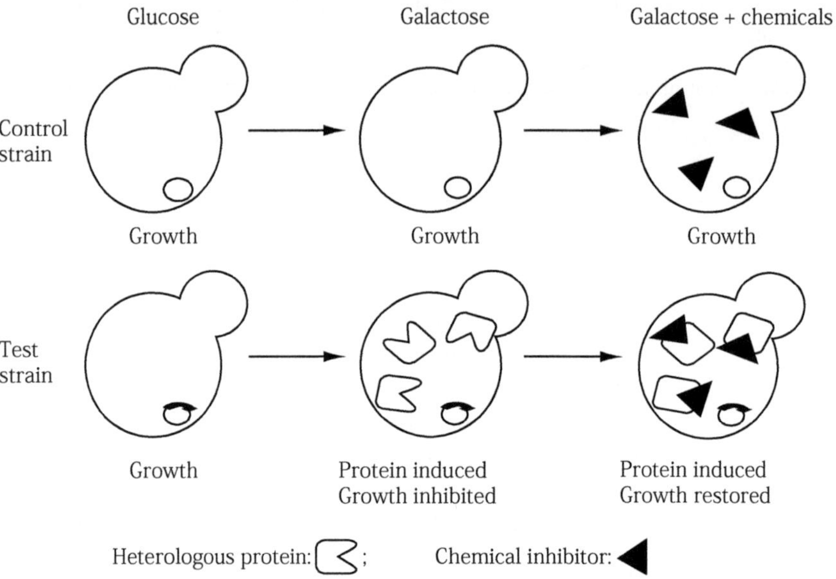

Fig. 1. Outline of the growth-restoration principle. Left column: a control strain containing an empty expression plasmid (*circle*) and a test strain containing an expression plasmid with the coding sequence of the gene of interest (*circle with curved arrow*) are grown in a medium containing glucose to repress expression. Middle column: the strains are transferred to a medium containing galactose to induce expression of the heterologous protein, resulting in inhibition of the growth of the test strain, but not of the control strain. Right column: the chemicals to be screened are added. Chemicals that inhibit the heterologous protein restore growth of the test strain and are scored as active.

plasmid containing the galactose-inducible *GAL1* promoter and introducing this construct into a yeast strain deleted of major drug-efflux pumps. The yeast strain bearing the plasmid is seeded at low density into multiwell plates in a medium containing galactose to induce protein expression; the chemicals to be screened are added, and yeast growth is measured by absorbance reading in a microplate reader. Active chemicals restore yeast growth. Simple controls can eliminate false-positive readings. Specialized secondary assays can then be used to verify the inhibitory activity of the active chemicals in the biological systems of interest. The screening assay is the same for different proteins. It is inexpensive, consists of very few steps, and its positive readout tends to exclude toxic compounds.

2. Materials

2.1. Yeast Strains and Expression Plasmids

1. Yeast strain EIS20-2B (*MAT**a**, ade2-1, his3-11, 15 leu2-3, 112 trp1-1, ura3-1, can1-100, pdr5Δ, snq2Δ)*, lacking the major efflux pumps Pdr5p and Snq2p *(1)* or strain FY1679-28C/TDEC (MAT**a**, *pdr1Δ2::TRP1, pdr3Δ::HIS3, ura3-52, leu2Δ, trp1-Δ63, his3Δ200, GAL2+*), lacking the drug efflux pump transcription factors Pdr1p and Pdr3p *(5)* (*see* **Note 1**).
2. Yeast expression plasmid: A multicopy plasmid for expression of proteins in yeast under the control of an inducible GAL1 promoter, such as pARC25B (LEU selection) *(1)*, or Gateway-compatible plasmids pJG482 (URA selection), pJG483 (LEU selection), or pJG514 (TRP selection) *(6)*. It is also possible to use plasmids for the expression of proteins fused to a V5-6xHis tag, such as pJG484 (LEU selection), pJG485 (URA selection), or pJG516 (TRP selection) *(6)* (*see* **Note 2**).

2.2. Yeast Transformation and Growth

1. Yeast extract peptone dextrose (YPD or YEPD) liquid medium: In a 200-mL glass bottle, weigh 1-g Bacto yeast extract (BioShop Canada, Inc., Burlington, ON) and 2-g Bactopeptone (BioShop), add 90-mL water, and autoclave the solution. When the autoclaved solution is cool, add 10 mL of 20% sterile glucose (*see* **step 10**). Store at room temperature.
2. 50% (w/v) polyethylene glycol 4000 (PEG): To a 150-mL beaker containing 30-mL water, add 50-g PEG (BDH, Inc., Toronto, ON). Stir with gentle heat to speed up dissolution of PEG, cool to room temperature, and adjust the volume to

100 mL with water. Transfer to a glass bottle. Autoclave and store at room temperature.

3. 1.0 M lithium acetate: Dissolve 10.2 g of lithium acetate (BDH) in 100-mL water. Autoclave the solution and store at room temperature.
4. 10 mg/mL sheared salmon sperm DNA (SSS DNA): Dissolve 10 mg of salmon sperm DNA (Sigma, St. Louis, MO; single-stranded carrier DNA) in 100 mL of TE (10 mM Tris–HCl, 1 mM disodium EDTA, pH 8.0) on a stir plate overnight at 4°C. Dispense 1.0 mL into 1.5-mL microcentrifuge screw cap tubes and store at –20°C. To denature the carrier DNA, place a tube in a boiling water bath for 5 min and chill immediately in an ice/water bath before use *(7)* (*see* **Note 3**).
5. Plasmid DNA or linear DNA, 1–2 μg: The DNA preparation should be quite pure.
6. Synthetic Complete (SC) mix: 0.6-g adenine, 0.6-g uracil, 0.6-g tryptophan, 0.6-g histidine, 0.6-g arginine, 0.6-g methionine, 0.9-g tyrosine, 0.9-g lysine, 1.5-g phenylalanine, 6.0-g threonine, 3.0-g aspartic acid, 1.8-g isoleucine, 4.5-g valine, 1.8-g leucine (Sigma). Weigh all ingredients, mix together, and use a mortar and pestle to grind into a homogeneous powder. Store in 50-mL Falcon tubes at room temperature.
7. SC dropout mix: Weigh all the ingredients except for leucine, tryptophan, or uracil, depending on the plasmid-selection marker used, and mix, grind, and store as in **step 6**.
8. SC selection medium: Yeast nitrogen base without amino acid (BD/Difco, Sparks, MD) 0.67%, SC dropout mix 0.067%, dissolve in water. Add 0.25 mL of 1N NaOH to every 100-mL medium to raise the pH of the solution to 6.5. Autoclave the solution and store at room temperature.
9. SC selection medium containing 2% agar and 2% glucose: Yeast nitrogen base without amino acid 0.67%, SC dropout mix 0.067%, dissolve in water. Add 0.25 mL of 1N NaOH and 2-g Bacto Agar (BioShop) to every 100 mL of medium. Autoclave the solution. When the autoclaved solution has cooled down to about 45°C, add 10 mL of 20% glucose (*see* **step 10**) to every 100 mL and pour 20 mL into sterile Petri dishes. Store the agar plates in a plastic bag at room temperature.
10. Glucose (BioShop): Dissolve 20 g in 100-mL warm water (20% solution), and filter sterilize using 0.22-μm filter into a sterile bottle. Store at room temperature.

11. Galactose (BioShop): Dissolve 20 g in 100-mL warm water (20% solution), and filter sterilize using 0.22-μm filter into a sterile bottle. Store at room temperature.
12. SC selection medium containing 2% glucose: Store in a sterile bottle at room temperature.
13. SC selection medium containing 2% galactose: Store in a sterile bottle at room temperature.
14. 50% Glycerol (glycerin; Fisher Scientific; Fairlawn, NJ). Dissolve 25-mL glycerol in 25-mL water. Autoclave the solution and store at room temperature.
15. Frozen stock of yeast strain: Mix 100 μL of 50% glycerol (v/v, autoclaved) sterile solution and 100 μL of yeast cell suspension in a sterile cryotube. Store the aliquot at −80°C.
16. Transformation mix: 240-μL PEG 4000 solution, 36-μL 1.0 M lithium acetate, 5-μL SSS DNA, 80-μL sterile water (total volume = 361 μL) for each transformation reaction.

2.3. Inhibitor Screening

1. Sterile nontissue culture 96-well plates with lids (Corning Costar, Lowell, MA).
2. Dispensing eight-channel pipettor (ePET, Biohit, Neptune, NJ).
3. Hand-pinning tool with 0.4- to 1.2-mm diameter pins (V&P Scientific, San Diego, CA) or automated pinning instrument such as TAS1 robot with a 0.7-mm diameter 96-pin tool (Bio-Robotics, Cambridge, England) (*see* **Note 4**).
4. 30°C incubator and temperature-maintained shaker incubator.
5. Humidifier box: Select a plastic box with lid that will hold the required number of plates and will fit into the incubator. Line the bottom of the plastic box with a sheet of paper towel and add sufficient water to wet the paper. Place the lid of a smaller plastic box on the paper towel to make a platform onto which you will put the 96-well plates so they do not touch the damp paper. Cover the humidifier box lightly with the lid to ensure air circulation.
6. Microplate reader and cuvette spectrophotometer with a 595- or 600-nm filter (e.g., Dynex Opsys MR, Chantilly, VA).
7. Chemical or natural product extract library: Compounds or extracts are dissolved in DMSO and stored in 96-well plates.

3. Methods

First, the coding sequence of the gene of interest is inserted into a yeast expression plasmid for galactose-inducible expression and introduced into yeast. The growth of this test strain in galactose is compared with the growth of a control strain containing the empty plasmid. If the heterologous protein causes growth inhibition, then a screen for chemicals that restore growth can be carried out. The chemicals showing most growth restoration are then tested in a concentration curve to confirm screening results, establish potency, and eliminate those also showing toxicity. The selected compounds can be further tested for selectivity using a strain expressing an unrelated protein that also causes growth inhibition. Follow-up assays must be used to confirm direct activity on the protein of interest.

3.1. Plasmid Preparation and Yeast Transformation [(7); see **Note 5**)]

1. Introduce the coding sequence of the gene of interest into the yeast expression plasmid using standard molecular biology techniques (*see* **Note 6**).
2. Grow the *S. cerevisiae* strain in 5-mL YPD liquid medium overnight in a shaker at 220 rpm and 30°C.
3. Measure the A_{600} of the overnight yeast suspension and dilute to an A_{600} of 0.5 in 20-mL YPD liquid medium. Incubate the cells for 4–5 h in an incubator shaker at 220 rpm and 30°C until the A_{600} reaches 2.0.
4. Transfer the cells to a centrifuge tube and centrifuge at 2000 × *g* for 8 min. Suspend the pellet with 20-mL sterile water and centrifuge again.
5. Resuspend the pellet in 1-mL sterile water and transfer to a 1.5-mL microfuge tube. Centrifuge at 16,000 × *g* for 2 min.
6. Resuspend the pellet in 0.5-mL sterile water. Transfer 100 μL into 1.5-mL microfuge tubes, one per transformation reaction.
7. Centrifuge at 16,000 × *g* for 2 min and discard the supernatant.
8. Prepare transformation mix.
9. Transfer 361-μL transformation mix to each 1.5-mL microfuge tube containing a yeast pellet.
10. Add 1–2 μL (0.5–1.0 μg) empty plasmid DNA or plasmid containing the gene of interest. If using recombinational cloning, add 100 μL of PCR-amplified gene-coding sequence containing the appropriate flanking DNA and 0.5–1.0 μg of linearized plasmid DNA *(1, 3)*. Gently suspend the yeast pellet by pipetting up and down (*see* **Note 6**).

11. Incubate the tubes in a water bath at 42°C for 40 min.
12. Centrifuge the tubes at 16,000 × *g* for 2 min and discard the supernatant.
13. Add 1-mL sterile water to each tube and mix the pellet gently by pipetting up and down.
14. Spread 50 or 100 μL of this suspension onto agar plates of appropriate SC selection medium containing 2% glucose but lacking uracil or tryptophan, depending on the plasmid used.
15. Incubate the plates at 30°C for 2–3 days until colonies appear.
16. Select 5–6 colonies per transformation to test for growth inhibition.

3.2. Test for Growth Inhibition

1. Inoculate the control strain with the empty plasmid and 5–6 colonies of the test strain bearing the plasmid with the gene of interest into tubes containing 2-mL SC selection medium with 2% glucose. Grow cells overnight at 30°C with shaking at 220 rpm.
2. The next day, transfer 1 mL of each overnight culture to microfuge tubes. Centrifuge at 4,700 × *g* for 5 min. Discard the supernatants, wash the pellets with sterile water, and centrifuge again at 4,700 × *g* for 5 min (*see* **Note 7**).
3. Resuspend the pellets in 1-mL sterile water, measure the A_{600}, and dilute the cells to A_{600} = 0.01 in appropriate SC liquid-selection medium containing 2% galactose.
4. Transfer 100 μL of yeast cells containing the control or test plasmid to wells of sterile 96-well plates, in several replicates.
5. Place the plates in the humidifier box and incubate them at 30°C for 40–42 h.
6. Shake the plate on a vortex at low speed (e.g., setting 4 of a Genie 2 Vortex mixer) for 90 s to resuspend the yeast cells and measure the A_{600} using a 96-well plate reader (*see* **Note 8**).
7. Calculate the percent growth inhibition by dividing the average A_{600} of transformants containing test plasmid by the A_{600} of transformants containing the empty plasmid, multiplying by 100, and subtracting this value from 100. A growth inhibition level of 50–80% is ideal for screening (*see* **Note 9**).
8. Prepare frozen stocks of selected strains.

3.3. Inhibitor Screening

1. The day before screening, inoculate the control strain containing the empty plasmid and the selected test strain bearing the plasmid with the gene of interest into 2 mL of SC selection medium containing 2% glucose. Grow cells overnight at 30°C with shaking at 220 rpm.
2. The next day, transfer 1 mL of overnight culture to a microfuge tube. Centrifuge at 4700 × *g* for 5 min. Discard the supernatant, wash the pellet with sterile water, and centrifuge at 4700 × *g* for 5 min, to eliminate traces of glucose.
3. Remove the plates containing chemicals to be tested from the freezer and thaw at room temperature for at least 30–60 min.
4. Suspend the pellet in 1-mL sterile water, measure the A_{600} and dilute the cells to A_{600} = 0.01 in appropriate SC liquid selection medium containing 2% galactose. Prepare ≥10 mL diluted test cells for each 96-well plate to be tested. A lower volume of control cells is required.
5. Transfer 100 μL of test cells to all but four wells of sterile 96-well plates using a dispensing eight-channel pipettor. Add 100-μL medium without yeast to two control wells and 100-μL medium with control cells to two wells (*see* **Note 10**).
6. Prepare a control 96-well plate. To columns 1–4 (32 wells) add 100-μL control cells diluted to A_{600} = 0.01. To columns 5–8 add 100-μL test cells diluted to A_{600} = 0.01, and to columns 9–12, add 100-μL medium without cells.
7. Transfer the chemicals from storage plates to plates containing yeast using a hand-held pinning tool or a robotic pinning tool. Clean and disinfect the pinning tool by dipping and shaking the pins in 10% bleach for 10 s, followed by dipping and shaking in 70% ethanol for 10 s, followed by air drying or drying over a flame. When pins are cool, dip the pinning tool into a chemical storage plate, remove the pinning tool carefully without touching the edges of the well, dip into the test plates without touching the edges of the wells, and remove in the same manner. Wash and disinfect the pins and repeat the process until all chemicals have been transferred to test plates (*see* **Note 11**).
8. Place the plates in the humidifier box and incubate them at 30°C for 40–42 h.
9. Shake stacks of five plates on a vortexer at low speed (e.g., setting 4 of a Genie 2 Vortex mixer) for 90 s to resuspend the yeast cells, and measure the A_{600} using a 96-well plate reader (*see* **Note 8**).

3.4. Growth Restoration Calculation

1. Control plate: Calculate the average A_{600} of test wells (columns 1–4), control wells (columns 5–8), and medium-only wells (columns 9–12).
2. Calculate the % growth restoration for each compound tested using the formula: % Growth restoration = (Test & chem – Test)/(Control – Test) × 100 *(3)*, where "Test & chem" is the A_{600} reading of a well containing the test strain treated with a chemical, "Test" is the average A_{600} of test cells not treated with chemicals determined from the control plate, and "Control" is the average A_{600} of control cells determined from the control plate.
3. Select as "actives" the wells showing highest levels of growth restoration (*see* **Note 12** and **Fig. 2**).

3.5. Active Chemical Confirmation

1. Visually inspect "active" wells in an inverted microscope to ensure that the increased A_{600} reading is indeed due to an increased number of yeast cells rather than being due to compound precipitation or contamination by other microorganisms (*see* **Note 13**).
2. To confirm the primary screening results, retest the activity of each active chemical at various concentrations against both the test and control strains. Establish the EC_{50} for each active compound and select compounds combining high potency and low toxicity (*see* **Note 14** and **Fig. 3**).
3. The active compounds can also be tested against a test strain for an unrelated gene that also causes growth inhibition when overexpressed in yeast. Chemicals that restore growth by general mechanisms, such as interference with the activity of the *GAL1* promoter, should also restore growth inhibited by any gene.

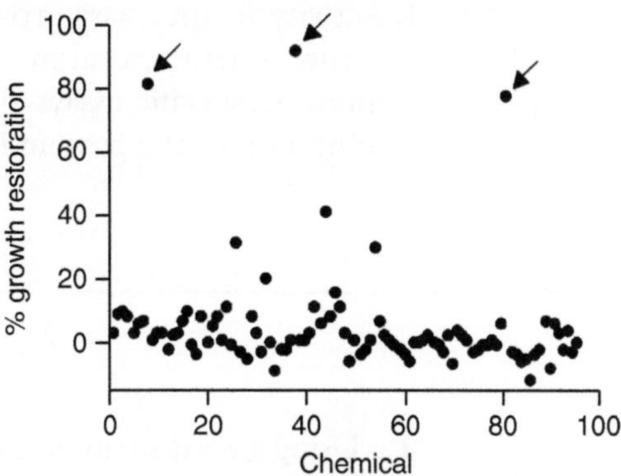

Fig. 2. Example of screening results obtained with one 96-well plate of chemicals. Most chemicals showed little or no growth restoration. Some caused growth restoration values below 0% because they are toxic to yeast. The three chemicals causing >50% growth restoration were scored as active (*arrows*).

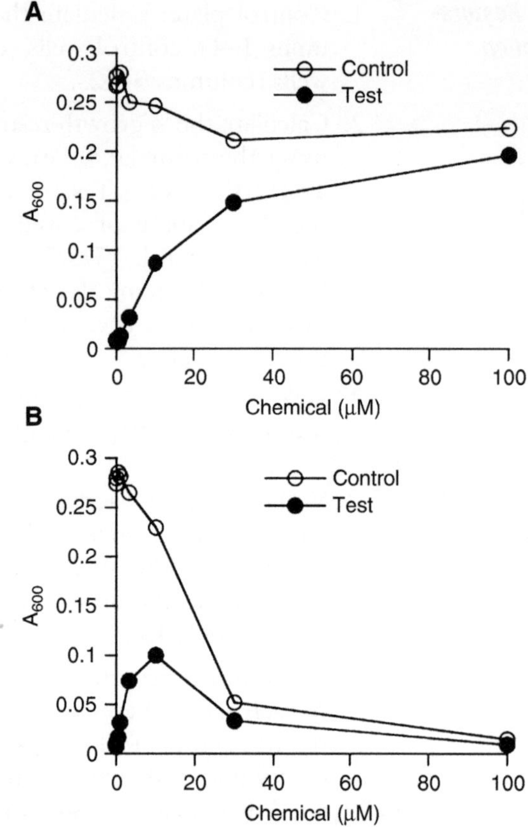

Fig. 3. Effects of two active chemicals on the growth of control and test strains. (**A**) The chemical causes concentration-dependent growth restoration of the test strain and shows little toxicity to the control strain. (**B**) The chemical causes growth restoration at low concentrations, but its activity is reduced at high concentrations. The chemical is toxic to the control strain at high concentrations. This chemical is less selective in yeast than the one shown in (**A**), and likely also in other biological systems.

4. Activity in the yeast growth restoration assay does not constitute a demonstration of direct inhibition of the protein of interest. Specific assays should be used to verify directly the inhibition of the protein in a suitable biological system.

4. Notes

1. Using a yeast strain deleted of one or more drug efflux pumps, instead of a wild-type strain, will increase the number of chemicals that can penetrate the yeast cell and increase the probability of finding inhibitors. A strain lacking *PDR5* and *SNQ2* has been used successfully in our laboratory. Strains deleted of

additional pumps are available, such as AD1-3⁻AD123 (*Mat*α, *pdr1-3, ura3, his1, yor1Δ::hisG, snq2Δ::hisG, pdr5Δ::hisG*) lacking four major drug efflux pumps and strain AD1-9⁻ (*Mat*α, *ura3, his1, yor1Δ::hisG, snq2Δ::hisG, pdr5Δ::hisG, pdr10Δ::hisG, pdr11Δ::hisG, ycf1Δ::hisG, pdr3Δ::hisG, pdr15Δ::hisG, pdr1Δ::hisG*) *(5)*. However, such strains may be less suitable for inhibitor screening because of their slow growth.

2. A number of expression plasmids are available, including Gateway-compatible ones. Start with a multicopy plasmid for higher protein expression. If growth inhibition is too strong, use a plasmid that can be integrated in a single copy at a chromosomal locus. Plasmids for protein expression fused to a V5-xHis tag can be useful to verify expression of the protein upon incubation in galactose.
3. Denatured carrier DNA can be boiled 3 or 4 times without loss of activity.
4. The volume of chemical that will be transferred, and hence the chemical concentration used in screening, will depend on the pin diameter and geometry, the concentration of the chemical, the depth of the solution, as well as the solvent used, usually DMSO. For example, our chemical plates contain 100 μL of 5-mM solutions, and our 0.7-mm diameter pin tool transfers 0.34 μL, for a final chemical concentration of 17 μM in the test plates. We have had success testing pure chemicals at 5–20 μM.
5. All procedures with yeast and sterile media are carried out on a benchtop near a lighted Bunsen gas burner that creates an updraft to help prevent contamination of the media or the yeast strain. A sterile hood is not necessary.
6. As an alternative to using restriction fragments and ligations, the coding sequence can be introduced into the plasmid by recombinational cloning in yeast *(1, 3)*. In addition, many sequence-verified human and mouse gene-coding sequences can be purchased in Gateway entry plasmids and transferred into Gateway destination vectors for yeast expression by in vitro recombination (http://www.invitrogen.com/content.cfm?pageid = 4072).
7. It is not sufficient to dilute cells grown in a medium containing glucose to a medium containing galactose, because the cells will use glucose preferentially over galactose, even when glucose is present in small amounts. This washing step is important to eliminate all traces of glucose from the medium.
8. The yeast will settle to the bottom of the wells during growth. For accurate A_{600} measurements, it is of paramount importance to resuspend them thoroughly before reading

the plates. The plates can be checked in an inverted microscope to ensure that suspension is complete.

9. If growth inhibition is very high (>90%), determine whether the cells are dead by adding 2% glucose and incubating for an additional 48 h. If the cells still do not grow, they are dead, so there is no point in trying to rescue growth with inhibitors. If cells are dead or growth inhibition is greater than 90%, consider integrating the plasmid at a chromosomal locus to maintain it in single copy instead of multiple copies when present as an episome. This will result in lower protein expression levels and, hopefully, lower growth inhibition. If growth inhibition is too low, consider changing the growth conditions *(4)*.
10. Place controls according to where empty wells are positioned in your chemical plates. If the chemical plates contain no empty wells, place controls in a separate plate, but run a control plate for each screening experiment. Screens can be carried out in duplicate, but we prefer to screen single points in order to assay larger numbers of compounds.
11. It is very important to wash and disinfect the pins thoroughly to prevent contaminating your chemical plates with yeast or with other chemicals. Always check that you use the same plate orientation for chemical plates and test plates (e.g., position A1 at the top left).
12. To select active chemicals for secondary assays, it is useful to graph the percent growth restoration obtained for many chemicals. We generally select as active chemicals all compounds showing >50% growth restoration, or obvious outliers (**Fig. 2**).
13. False-positive readings can be caused by compound precipitation, which can obstruct the light path. Compound precipitation is easily seen in an inverted microscope as crystals or powdery deposits at the bottom of the well. Green-colored chemicals can also provide false-positive readings. Contamination by other microorganisms is possible, but we have never observed it in our experiments.
14. The shape of the concentration dependence of growth restoration provides both an indication of the potency of the compound for growth restoration and of its toxicity at higher concentrations. Compounds that cause a high extent of growth restoration at low concentrations and that do not inhibit the growth of the control or test strains (**Fig. 3A**) are more likely to show selectivity in your biological assay than compounds that inhibit growth at concentrations barely higher than active concentrations in the growth restoration assay (**Fig. 3B**).

Acknowledgments

The author would like to thank Phil Hieter, Stuart Tugendreich and Teri Melese for their advice. This work was supported by grant 017392 from the National Cancer Institute of Canada, with funds from the Canadian Cancer Society.

References

1. Tugendreich, S., Perkins, E., Couto, J., Barthmeier, P., Sun, D., Tulac, S., et al. (2001) A streamlined process to phenotypically profile heterologous cDNAs in parallel using yeast cell-based assays. *Genome Res.* 11, 1899–1912.
2. Kurtz, S., Luo, G., Hahnenberger, K. M., Brooks, C., Gecha, O., Ingalls, K., et al. (1995) Growth impairment resulting from expression of influenza virus M2 protein in *Saccharomyces cerevisiae*: identification of a novel inhibitor of influenza virus. *Antimicrob. Agents Chemother.* 39, 2204–2209.
3. Perkins, E., Sun, D., Nguyen, A., Tulac, S., Francesco, M., Tavana, H., et al. (2001) Novel inhibitors of poly(ADP-ribose) polymerase/ PARP1 and PARP2 identified using a cell-based screen in yeast. *Cancer Res.* 61, 4175–4183.
4. Vottero, E., Balgi, A., Woods, K., Tugendreich, S., Melese, T., Andersen, R. J., et al. (2006) Inhibitors of human indoleamine 2,3-dioxygenase identified with a target-based screen in yeast. *Biotechnol. J.* 1, 282–288.
5. Rogers, B., Decottignies, A., Kolaczkowski, M., Carvajal, E., Balzi, E., and Goffeau, A. (2001) The pleiotropic drug ABC transporters from *Saccharomyces cerevisiae*. *J. Mol. Microbiol. Biotechnol.* 3, 207–214.
6. Geiser, J. R. (2005) Recombinational cloning vectors for regulated expression in *Saccharomyces cerevisiae*. *BioTechniques* 38, 378–389.
7. Gietz, D. and Woods, R. (2006) Yeast transformation by the Li/SS carrier DNA/PEG method, in Methods in Molecular Biology: Yeast Protocols (Xiao, W., ed.). Humana Press, Totowa, NJ, pp. 107–120

Chapter 10

Assay for Isolation of Inhibitors of Her2-Kinase Expression

Gabriela Chiosis and Adam B. Keeton

Summary

Her2 (ErbB2) protein is overexpressed in breast and other solid tumors, and its expression is associated with progressive disease. Current therapies directed toward Her2 either block dimerization of the receptor or inhibit tyrosine kinase activity to disrupt intracellular signaling. However, little is known about alternative mechanisms for suppressing Her2 expression, possibly by inducing degradation or blocking synthesis. Here, we describe a hybrid western-blotting and enzyme-linked immunosorbent assay (ELISA) designed to identify in low- to medium-throughput format noncytotoxic compounds that reduce expression of Her2 protein.

Key words: Breast cancer, Heat-shock protein 90 (Hsp90), Her2 (ErbB2) tyrosine kinase, Whole-cell immunoblot.

1. Introduction

Human cancers frequently express high levels of transmembrane tyrosine kinases of the Her family, comprising four closely related transmembrane receptor tyrosine kinases, Her1-4 *(1–3)*. Dysregulation of the activity of these receptors often leads to cellular transformation and cancer. These receptors act as a layered network, in which the four receptors are activated by ten growth-factor ligands of the EGF/neuregulin family, to bring about formation of homo- and heterodimers with different signaling capabilities. The first level of regulation of the network relates to formation of the primary signaling unit, which depends on the proper temporal and spatial coexpression of a ligand and two receptors. The major signaling pathways activated by this signaling unit of the Her-receptors include the Ras-Raf1-Mek-Erk, and the PI3K-PDK1-Akt pathways,

P.A. Clemons et al. (eds.), *Cell-Based Assays for High-Throughput Screening, Methods in Molecular Biology, Vol. 486*

DOI: 10.1007/978-1-60327-545-3_1

both of which culminate in activation of transcription programs, as well as cyclin-dependent kinases, leading to progression through the cell cycle (**Fig. 1**). While Her1 (EGFR), Her3, and Her4 each bind multiple ligands, Her2 (ErbB2, Neu) is a ligandless receptor that enhances and prolongs signaling upon heterodimerization with other Her-receptors. Its overexpression has been detected in up to 30% of breast and ovarian cancers, but also in other types of cancers including lung, gastric, and oral cancers, and has been correlated with invasive and poor prognostic features *(1–3)*. In addition, Her2 has been implicated in mediating increased resistance to chemotherapeutic agents. Current drugs targeted at Her2 receptors work either by blocking receptor dimerization (antireceptor antibodies such as Herceptin) or by inhibiting tyrosine kinase activity (TK inhibitors) *(4, 5)*. Identification of novel means to regulate the expression of this kinase should provide additional clinical opportunities for successfully interdicting signaling through Her2-containing receptor complexes.

The stability of Her2 is regulated by the chaperone Hsp90, suggesting inhibition of Hsp90 as an alternative way to block the activity of the kinase *(6)*. Hsp90 activity can be inhibited with agents that bind to its N-terminal region ATP/ADP pocket, such

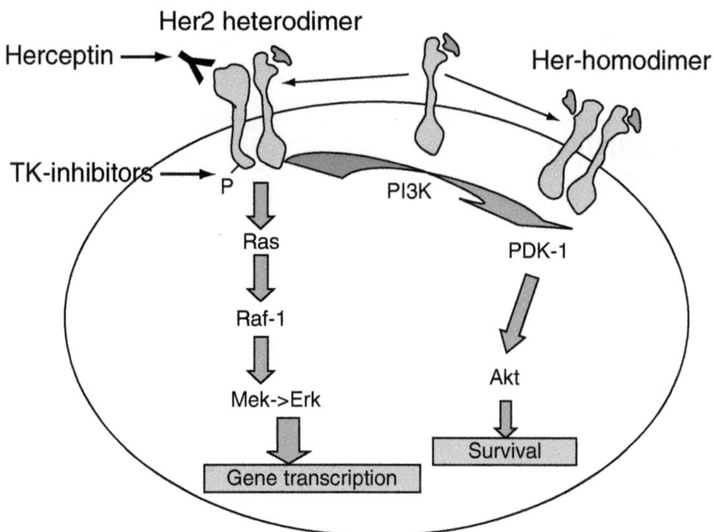

Fig. 1. Ligand binding to inactive monomeric Her receptors is followed by dimer formation and kinase activation. Upon recruitment into heterodimers, the ligandless coreceptor Her2 enhances and prolongs signaling through several pathways, including the indicated mitogen-activated protein kinase (MAPK) pathway and the phosphatidylinositol 3-kinase (PI3K) route, leading to activation of the Akt/PKB kinase. Points of intervention by current drugs targeting Her2 receptors (Herceptin, tyrosine kinase inhibitors) are indicated. The abbreviations used are as follows: *PDK-1* phosphoinositide-dependent kinase 1, *TK* tyrosine kinase.

as geldanamycin (GM) and radicicol (RD). Addition of these agents to cells induces the proteasomal degradation of Her2 *(7)*. HDAC inhibitors that acetylate Hsp90 also cause Her2 degradation *(8)*. Thus, agents that target Hsp90 through diverse mechanisms can alter Her2 expression.

An assay designed to identify or evaluate compounds that reduce Her2 levels is represented later (*see* **Note 1**). This is a low- to medium-throughput *cytoblot-type* assay based on a nonhomogeneous ELISA-type readout that detects endogenous cellular Her2 levels *(9)* and has been validated by us in the evaluation of Hsp90 inhibitors (approximately 300 compounds) *(10, 11)*. In principle, the assay may test compounds that affect kinase transcription, translation, or stability.

2. Materials

2.1. Cell culture and Compound Dispersion

1. The human cancer cell line SKBr3 is maintained in 1:1 mixture of DME:F12 supplemented with 2 mM glutamine, 50 units/mL penicillin, 50 units/mL streptomycin, and 5% heat inactivated fetal bovine serum (FBS) (Gemini Bioproducts, West Sacramento, CA), and incubated at 37°C in 5% CO_2. Stock cultures are grown in T-175 flasks containing 30 mL of DME (HG, F-12, nonessential amino acids, penicillin, and streptomycin), glutamine, and 10% FBS.
2. Cells are dissociated with 0.05% trypsin and 0.02% EDTA in phosphate-buffered saline (PBS) without calcium and magnesium to provide experimental cultures.
3. Using a multichannel pipettor, SKBr3 breast cancer cells are plated at 3,000 cells/well in 100 μL of growth medium in the 96-well microtiter plates, and allowed to attach for 24 h at 37°C and 5% CO_2 (*see* **Note 2**). Some wells are left without cells to serve as the blank control.
4. Clear-bottom 96-well microtiter plates (Corning Incorporated, Lowell, MA).
5. Multichannel pipettors.
6. SKBr-3 cells (American Type Culture Collection, Manassas, VA) (*see* **Note 3**).
7. Dulbecco's modified Eagle's medium (DMEM) with Ham's nutrient mixture F12 (DMEM/F12) supplemented with 30 mM d-glucose and nonessential amino acids (Gibco/Invitrogen, Carlsbad, CA), and 10% fetal bovine serum (FBS, Gemini Bio-Products) and Pen-Strep solution (10,000 U/mL penicillin G and 10 mg/mL streptomycin; Gemini Bio-Products).

8. Solution of trypsin (0.05%) and ethylenediamine tetraacetic acid (EDTA; 0.02%) in PBS (Gibco/Invitrogen).
9. Compound library plated in microtiter plates and dissolved at a known concentration in DMSO (Sigma-Aldrich, St. Louis, MO) (*see* **Note 4**).
10. Positive control: an Hsp90 inhibitor such as radicicol (Sigma-Aldrich) or geldanamycin (Sigma-Aldrich) (*see* **Note 5**).

2.2. Assay Development and Detection

1. Vacuum source attached to an 8-channel aspirator used to remove liquid from the microplates.
2. Pipet basin (ThermoFisher, Waltham, MA).
3. Ice-cold Tris-buffered saline (TBS) containing 25 mM Tris–HCL, 0.13 M NaCl, 0.0027 M KCl (Fisher Scientific, Fair Lawn, NJ) supplemented with 0.1% Tween-20 (Fisher Scientific, Pittsburgh, PA) (*see* **Note 6**).
4. Methanol, chilled to –20°C (Fisher Scientific) (*see* **Note 7**).
5. Antibodies: anti-Her2 (c-erbB-2) antibody (Zymed Laboratories/Invitrogen), goat anti-rabbit HRP-linked antibody (Santa Cruz Biotech, Santa Cruz, CA), and normal rabbit IgG (Santa Cruz) (*see* **Note 8**).
6. ECL Western Blotting Detection Reagent (GE Healthcare, Piscataway, NJ).
7. Blocking solution, SuperBlock (Pierce Biotechnology, Rockford, IL) (*see* **Note 9**).
8. Bicinchoninic acid reagent (BCA; Pierce).
9. Standard solution of bovine serum albumin (BSA; Pierce).
10. Rocking platform.
11. Analyst GT plate reader (Molecular Devices, Sunnyvale, CA) (*see* **Note 10**).

3. Methods

The identification of compounds that alter protein levels has required cumbersome in-vitro analyses. This procedure involves treating cultured cells with individual drugs, followed by detergent lysis, polyacrylamide gel electrophoresis of total cellular proteins, and Western blotting to determine protein levels. This methodology is decidedly unsuitable for rapid compound analysis. In contrast, the Her2-blot, which relies on whole-cell immunodetection of the desired proteins, utilizes a minimal number of cells, yet it is sufficiently sensitive and reproducible to permit quantitative determinations. The assay is a hybrid of a Western blot and an enzyme-linked immunosorbent assay (ELISA), and is a modified version of the cytoblot assay developed by Stockwell et al. *(12)*. The cytoblot assay is adaptable for screening numerous protein modifications because the only requirement is a specific primary antibody directed against the protein of interest. It has been applied to HTS to identify many bioactive small molecules, such as inhibitors of mitotic spindle bipolarity *(13)*, mitosis modulators *(14)*, inhibitors of phospho-STAT3 *(15)*, inhibitors of HDAC6-mediated tubulin deacetylation *(16)*, and modulators of the human chromatid decatenation checkpoint *(17)*. Our contribution to the cytoblot is that for the first time we detected a specific decrease in the amount of a protein *(9)*. Testing for the presence of a gene product is far simpler than testing for the absence or near absence of a gene product, because the amount of protein presumably increases with time in the first instance, and thus distinguishes itself from the background. In contrast, when the disappearance of a product is measured, the best resolution from background is at time zero. Notwithstanding this theoretical difficulty, meaningful measurements of Her2 can be obtained.

Figure 2 shows a schematic representation of the assay. The method consists of plating cells in microtiter dishes and treating wells with small molecules at equal concentrations, to identify agents that alter cellular levels of the protein, or at a concentration range to determine their potency in degrading Her2. Following treatment, cells are fixed and permeabilized with methanol. An antibody against the protein and a secondary antibody linked to horseradish peroxidase (HRP) are added. Upon addition of a luminescent substrate, the signal emitted is read in a luminom-

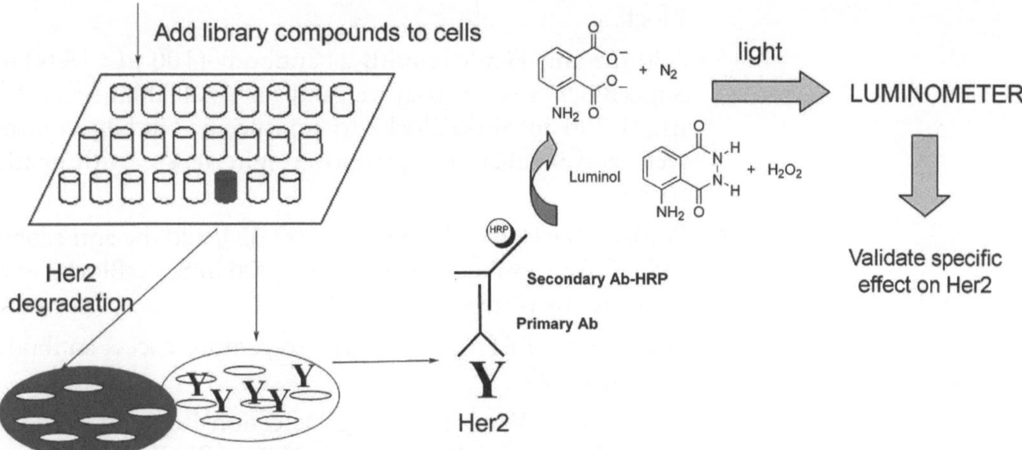

Fig. 2. Schematic of a Her2-blot.

eter. In an assay such as this, where a reduction in signal represents a *hit*, there are potential sources for false positives, such as from cytotoxic agents, which induce a decrease in the signal by cell death rather than through Her2 modulation. Thus, during screening it is important to identify cytotoxic compounds in the library that may reduce total cell number, rather than having a specific effect on Her2. For this reason, the assay can be multiplexed with an indicator of cell viability. A low-cost alternative to viability assays is to determine the amount of total protein in each well using the bicinchoninic acid (BCA) assay.

The Her2-assay described later is designed to include the following controls: vehicle-treated wells (control), wells containing cells treated with geldanamycin or radicicol (positive control; the signal should reach background levels), and wells containing only detection reagent (set as blank in SoftMaxPro software; the values will be subtracted from all analyzed wells). To control for nonspecific antibody binding to the plate, a normal rabbit IgG is added to the plate (background).

1. Dispense the compound library or vehicle (DMSO) to wells, and incubate plates for 12 h (or 6 h) at 37°C and 5% CO_2 (*see* **Note 11**). Use two wells for the positive control. For geldanamycin and radicicol, the final concentration of the compound in the well is 50 nM.

2. Remove media through aspiration and wash each well with ice-cold TBST (2 × 200 μL). Use a vacuum source (such as house vacuum) attached to an eight-channel aspirator to remove the liquid from plates.

3. Add cold methanol (100 μL at –20°C) to each well, and incubate the plate at 4°C for 10 min.

4. Wash each well with TBST (2 × 200 μL) to remove the methanol, and incubate at RT for 2 h with 200-μL SuperBlock.

5. Add the anti-Her-2 (c-erbB-2) antibody (100 μL; 1:200 in SuperBlock) to each assay well and a normal rabbit IgG (100 μL; 1:200 in SuperBlock) to nonspecific-binding control wells and incubate the plate overnight at 4°C with gentle rocking.

6. Wash each well with TBST (2 × 200 μL), add the anti-rabbit HRP-linked antibody (100 μL; 1:2,000 in SuperBlock), and incubate the plate at RT for 2 h.

7. Wash with TBST (3 × 200 μL) to remove excess antibody (*see* **Note 12**).

8. Add ECL™ Western Blotting Detection Reagents 1 and 2 in a 1:1 mixture (100 μL) (*see* **Note 13**). Read the plate

immediately in an Analyst GT plate reader. Scan each well for 0.1 s. Luminescence height = 0.6 mm; maximum integration = 100,000 μs.

9. Settings: readings per well = 1; target CV per well = 1.0%; raw data units = counts/sec; attenuation mode = out; motion settling time = 15 ms; luminescence aperture = 384/96 aperture; max cps = 100,046.9922 cps; min counts = 349 counts.
10. Import and analyze data in SoftMaxPro (Molecular Devices). Subtract readings from IgG control wells from all measured values. Calculate % Her2 level as luminescence readings resulting from drug-treated cells/untreated cells (vehicle treated) × 100 *(9)*. **Fig. 3** shows representative raw data obtained from a Her2-blot (*see* **Note 14**).
11. To determine the protein content in each well, wash plates from Her2-blot readings with TBST (2 × 200 μL) and incubate them with the BCA reagent (150 μL) for 30 min at 37°C. Add a standard solution of BSA to the blank wells.
12. Read absorbance using the Analyst GT (Molecular Devices). Instrument settings: read sequence = row; mode sequence = well; detection mode = a9; excitation side = top; emission side = top; lamp = continuous; top dichroic = 50:50 beam splitter; readings per well = 1; integration time = 100,000 μs; motion settling time = 200 ms; *Z* height = 5.33 mm, middle; excitation filter = fluorescence 530 nm. Import and analyze data in SoftMaxPro (Molecular Devices) and determine the protein content of each well (*see* **Note 15**).

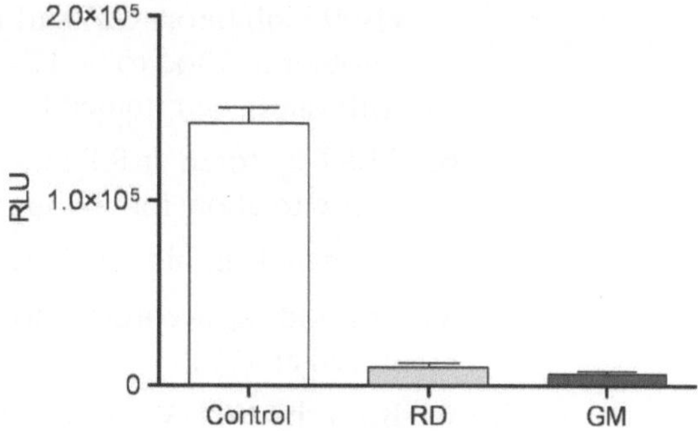

Fig. 3. Representative results of a Her2 blot. Readings obtained in wells containing SKBr3 cells treated for 12 h with DMSO only (control) (*average* 141715.3; *s.d.* 8573.8), 50 nM radicicol (RD) (*average* 9815.3; *s.d.* 2459.1), and 50 nM geldanamycin (GM) (*average* 6467.4; *s.d.* 1256.7), respectively.

4. Notes

1. The assay uses cells, media, buffers, and antibodies that are commercially available, and bulk amounts of these reagents can be purchased at reduced costs. Used at a 1:200 dilution, approximately 45-μL antibody is used for one 96-well plate at 100-μL assay solution/well. At retail price, one vial of antibody ($128 for 1 mL) would be sufficient to test over twenty 96-well plates, bringing the cost to roughly $5.50/ plate.
2. These conditions were optimized for the Analyst plate-reader and the stock of SKBr3 available in our lab. However, considering the variability among cell stocks and the different sensitivity of luminometers, it is highly recommended to perform cell-number optimizations as described in *(9)*. A 48-h attachment period may result in much better retention of cells during washes, thus lower well-to-well variability. If removing the liquid from the microtiter well plates manually using an eight-channel aspirator, position the pipet tips to the leftmost corner of each well to avoid accidental removal of attached cells. Cells are attached quite firmly, but pipetting should be done gently. This is usually achieved by letting liquid fall down the sides of wells, rather than directly into them.
3. SKBr3 breast cancer cells overexpress Her2, facilitating its detection in the cytoblot format.
4. DMSO levels in the final assay volume should not exceed 0.7% (v/v) because the organic solvent is toxic to SKBr3 cells.
5. The stability of Her2 is dependent on the Hsp90 protein chaperone system, and Hsp90 inhibitors induce the degradation of Her2 *(6, 7)*. IC_{50} values for the commercially available Hsp90 inhibitors GM and RD were determined in the Her2-cytoblot method to be 17 and 28 nM, respectively, consistent with values determined by Western blot analyses *(9)*.
6. TBST is stored at RT but placed on ice at least 30 min prior to use to allow for chilling.
7. Methanol should be cooled to –20°C before use.
8. Antibodies, according to manufacturers' instructions, are stored at 4°C.
9. Both the ECL Western Blotting Detection Reagents (1 and 2) and the Superblock are stored at 4°C.
10. As of 2007, www.globalspec.com lists 46 companies that make microplate readers. The choice for selecting a certain instrument may be determined by the assay type to be performed, the desired detection sensitivity, and by the available funds. Most screening facilities have several multimode plate readers

that can handle 96-well, 384-well, or 1536-well assay plates and support multiplate operations, either through stackers or robotic integration. Perkin Elmer (www.perkinelmer.com) offers several multilabel integratable plate readers for HTS. Among these are VICTOR3™, EnVision™, and ViewLux™. VICTOR3 is a multilabel, upgradable plate reader for fluorescence (top and bottom), luminescence, absorbance, UV absorbance, time-resolved fluorimetry, and fluorescence polarization detection technologies. A small benchtop unit, VICTOR3 operates as a stand-alone instrument or integrated into a robotic system. EnVision™ is a fast, sensitive, multitechnology plate reader designed for fast-paced HTS labs running many different types of assays. The EnVision employs application-specific optical modules, which enables easy customization to meet each user's specifications and needs. The ViewLux is an ultrahigh-throughput microplate imager for lead identification using any of the major assay technologies – FP, FI, TRF, luminescence, radiometric, and absorbance. Molecular Devices (www.moleculardevices.com) also offers several sensitive and versatile multimode readers. Among these are Analyst GT and Analyst HT instruments, which offer detection based on FI, FP, and time-resolved fluorescence resonance energy transfer (TR-FRET), including HTRF, luminescence, and absorbance (epi- and trans-). Each assay detection mode has been optimized to deliver enhanced performance in terms of sensitivity, dynamic range, signal-to-noise, and cross-talk. Analyst instruments support 96-, 384-, and 1536-well plate formats and are optimized for a good integration into robotic environments. Tecan (www.tecan.com) offers the Infinite™ F500 multidetection microplate reader. The Infinite™ F500 is a sensitive filter-based detection platform capable of reading 6- to 1,536-well microplates, including standard- and low-volume 384-well plates. This system supports a wide variety of measurement modes, including FI, absorbance, luminescence (flash and glow), FP, FRET and bioluminescence resonance energy transfer (BRET™), TRF, and TR-FRET-based assays. BioTek Instruments (www.biotek.com) has developed the Synergy™ HT Multi-Detection microplate reader that is compatible with most automated systems. This microplate reader for fluorescence, absorbance, and luminescence measurements was designed with an emphasis on superior performance in all detection methods. The Synergy HT has a 4-Zone™ temperature control system that ensures temperature uniformity necessary for kinetic assays.

11. It was determined that addition of an Hsp90 inhibitor to SKBr3 cancer cells induces the rapid proteasomal degradation of Her2, with most protein being depleted at 6 h. To observe maximal effects, a 12-h drug treatment is considered

more appropriate. The time frame, however, can be set to an interval that best suits the purpose of the screening assay. However, one should bear in mind that significantly longer treatment periods may increase the potential for cytotoxicity as an artifact in this assay.

12. It is critical to make sure that residual secondary antibody is completely washed out before adding ECL reagents. Any nonspecific residue will create outliers among the data.
13. For optimal reading results, when adding the ECL Western Blotting Reagents, it is recommended to remove significantly sized air bubbles. These can be gently popped using pipet tips or needles. This is a tedious process, and it is best to take great care to avoid the introduction of bubbles to begin with. Depending on the detection equipment, bubbles may drastically reduce signal intensity.
14. All *hit* compounds must be rescreened. Dose-dependent measurements should be conducted and IC_{50} values calculated for each *hit* to validate and rank the *hits* generated in the primary screen.
15. Alternatively, putative hits may be retested for cytotoxicity using the sulforhodamine B (SRB) colorimetric assay as described in the protocol (www.sigmaaldrich.com/sigma/bulletin/tox6bul.pdf). The assay is used for cell-density determination, based on the measurement of cellular protein content. The results are linear over a 20-fold range of cell numbers and the sensitivity is comparable to those of fluorometric methods; however, the method requires only simple equipment and inexpensive reagents.

Acknowledgments

The author would like to thank Joungnam Kim and Cristina C. Clement for technical assistance. This work was supported by R03 MH076408, and a Susan G. Komen grant, R01-SGK-BCTR0504381.

References

1. Yarden, Y. and Sliwkowski, M. X. (2001) Untangling the ErbB signalling network. *Nat. Rev. Mol. Cell. Biol.* 2, 127–137.
2. Klapper, L. N., Kirschbaum, M. H., Sela, M., and Yarden, Y. (2000) Biochemical and clinical implications of the ErbB/HER signaling network of growth factor receptors. *Adv. Cancer Res.* 77, 25–79.
3. Yu, D. and Hung, M. C. (2000) Overexpression of ErbB2 in cancer and ErbB2-targeting strategies. *Oncogene* 19, 6115–6121.
4. Ross, J. S. and Gray, G. S. (2003) Targeted therapy for cancer: the HER-2/neu and Herceptin story. *Clin. Leadersh. Manag. Rev.* 17, 333–340.

5. Arteaga, C. (2003) Targeting HERl/EGFR: a molecular approach to cancer therapy. *Semin. Oncol.* 3 Suppl 7, 3–14.
6. Citri, A., Gan, J., Mosesson, Y., Vereb, G., Szollosi, J., and Yarden, Y. (2004) Hsp90 restrains ErbB-2/HER2 signalling by limiting heterodimer formation. *EMBO Rep.* 12, 1165–1170.
7. Citri, A., Kochupurakkal, B. S., and Yarden, Y. (2004) The achilles heel of ErbB-2/HER2: regulation by the Hsp90 chaperone machine and potential for pharmacological intervention. *Cell Cycle* 1, 51–60.
8. Aoyagi, S. and Archer, T. K. (2005) Modulating molecular chaperone Hsp90 functions through reversible acetylation. *Trends. Cell Biol.* 11, 565–567.
9. Huezo, H., Vilenchik, M., Rosen, N., and Chiosis, G. (2003) Microtiter cell-based assay for the detection of agents that alter cellular levels of Her2 and EGFR. *Chem. Biol.* 10, 629–634.
10. Llauger, L., He, H., Kim, J., Aguirre, J., Rosen, N., Peters, U., Davies, P., and Chiosis, G. (2005) 8-Arylsulfanyl and 8-arylsulfoxyl adenine derivatives as inhibitors of the heat shock protein 90. *J. Med. Chem.* 48, 2892–2905.
11. He, H., Zatorska, D., Kim, J., Aguirre, J., Llauger, L., She, Y., Wu, N., Immormino, R. M., Gewirth, D. T., and Chiosis, G. (2006) Identification of potent water-soluble purine-scaffold inhibitors of the heat shock protein 90. *J. Med. Chem.* 49, 381–390.
12. Stockwell, B. R., Haggarty, S. J., and Schreiber, S. L. (1999) High-throughput screening of small molecules in miniaturized mammalian cell-based assays involving post-translational modifications. *Chem. Biol.* 2, 71–83.
13. Mayer, T. U., Kapoor, T. M., Haggarty, S. J., King, R. W., Schreiber, S. L., and Mitchison, T. J. (1999) Small molecule inhibitor of mitotic spindle bipolarity identified in a phenotype-based screen. *Science* 286, 971–974.
14. Haggarty, S. J., Mayer, T. U., Miyamoto, D. T., Fathi, R., King, R. W., Mitchison, T. J., and Schreiber, S. L. (2000) Dissecting cellular processes using small molecules: identification of colchicine-like, taxol-like and other small molecules that perturb mitosis. *Chem. Biol.* 4, 275–286.
15. Blaskovich, M. A., Sun, J., Cantor, A., Turkson, J., Jove, R., and Sebti, S. M. (2003) Discovery of JSI-124 (cucurbitacin I), a selective Janus kinase/signal transducer and activator of transcription 3 signaling pathway inhibitor with potent antitumor activity against human and murine cancer cells in mice. *Cancer Res.* 63, 1270–1279.
16. Haggarty, S. J., Koeller, K. M., Wong, J. C., Grozinger, C. M., and Schreiber, S. L. (2003) Domain-selective small molecule inhibitor of HDAC6-mediated tubulin deacetylation. *Proc. Natl. Acad. Sci. USA* 100, 4389–4394.
17. Haggarty, S. J., Koeller, K. M., Kau, T. R., Silver, P. A., Roberge, M., and Schreiber, S. L. (2003) Small molecule modulation of the human chromatid decatenation checkpoint. *Chem. Biol.* 12, 1267–1279.

Chapter 11

High-Content Screening: Flow Cytometry Analysis

Bruce S. Edwards, Susan M. Young, Irena Ivnitsky-Steele, Richard D. Ye, Eric R. Prossnitz, and Larry A. Sklar

Summary

The HyperCyt® high-throughput (HT) flow cytometry sampling platform uses a peristaltic pump, in combination with an autosampler, and a novel approach to data collection, to circumvent time-delay bottlenecks of conventional flow cytometry. This approach also dramatically reduces the amount of sample aspirated for each analysis, typically requiring ~2 µL per sample while making quantitative fluorescence measurements of 40 or more samples per minute with thousands to tens of thousands of cells in each sample. Here, we describe a simple robust screening assay that exploits the high-content measurement capabilities of the flow cytometer to simultaneously probe the binding of test compounds to two different receptors in a common assay volume, a duplex assay format. The ability of the flow cytometer to distinguish cell-bound from free fluorophore is also exploited to eliminate wash steps during assay setup. HT flow cytometry with this assay has allowed efficient screening of tens of thousands of small molecules from the NIH Small-Molecule Repository to identify selective ligands for two related G-protein-coupled receptors, the formylpeptide receptor and formylpeptide receptor-like 1.

Key words: Flow cytometry, Fluorescence, Formylpeptide receptor, Formylpeptide receptor-like 1, High-content screening, High-throughput screening, NIH Molecular Libraries Screening Center Network.

1. Introduction

Because of its facility for making high-content measurements, flow cytometry has played an accessory role in many phases of drug discovery in pharmaceutical and many biotechnology companies. Historically, it has been severely limited in throughput rates for the serial analysis of multiple discrete cell suspensions, with maximum speeds approaching only six samples per minute in the most advanced commercial system available at this writing. One feature

P.A. Clemons et al. (eds.), *Cell-Based Assays for High-Throughput Screening, Methods in Molecular Biology, Vol. 486*

DOI: 10.1007/978-1-60327-545-3_1

of conventional flow cytometry that hampers throughput is the one-sample-one-file paradigm – the practice of creating a separate data file for each cell suspension analyzed. Repetitive time delays are incurred for each sample that include one at the onset, while hardware and software are initialized for data collection, and another at the end, while data are saved to a file on a hard disk or other storage destination. A second rate-limiting feature is a consequence of the traditional cell sample-processing loop. The entire sample uptake/transport pathway is first rinsed with fluid to flush away the previous sample, and then the new sample suspension must be transported all the way from the source well to the point of analysis before the next data-acquisition sequence can be started.

A recently developed high-throughput (HT) flow cytometry sampling platform, designated HyperCyt® *(1)*, addresses these limitations. In contrast to the traditional sample-handling paradigm, the HyperCyt® HT flow cytometry approach is to fill the sample uptake/transport line with a stream of discrete sample particle suspensions, each typically ~2 μL aspirated from one of the source wells and separated from each other by air bubbles. The entire sample stream is continuously delivered to the flow cytometer, so that data from all the samples in a plate are acquired and stored in a single data file. A high-resolution time parameter is also recorded during data acquisition. Temporal gaps in particle detection are created in the data stream by the passage of the air bubbles, allowing the individual particle suspensions to be easily distinguished and separately evaluated when plotted in conjunction with the time parameter. This approach not only eliminates time lags associated with multiple rounds of initialization, data storage, and sample line filling, but also dramatically reduces the amount of sample aspirated for each analysis. Sampling rates in excess of one sample per second have successfully been demonstrated for endpoint assays, but a rate of ~40 samples/min is more routinely used as the most practical setting for producing robust assay analysis results. It is possible to make quantitative fluorescence measurements of thousands to tens of thousands of cells in each sample under such conditions of serial sample throughput *(2–5)*.

Described herein is a simple robust assay that detects active compounds on the basis of their ability to block the binding of a high-affinity fluorescent ligand to cellular receptors *(3–5)*. In negative-control wells, or in wells containing inactive compounds, cells exhibit bright green fluorescence due to the bound fluorescent ligand. Compounds with receptor-binding activity promote displacement of fluorescent ligand from the receptor, causing a decrease in green cell fluorescence intensity. Since ligand binding is measured directly, the assay detects active compounds independently of potential complexities in the patterns of the physiological responses of cells. For example, it should detect molecules regardless of whether they act as agonists or antagonists or mediate full, partial, or selective (e.g., signaling pathway-specific) activity. This is an advantage over

HT functional-response assays (e.g., intracellular Ca^{2+} flux), which are typically designed to detect exclusively agonists or antagonists, and are less sensitive for detecting compounds with selective or partial response activity. The assay is homogeneous, in that cells, compounds, and fluorescent peptide are added in sequence, and the wells subsequently analyzed without intervening wash steps. Elimination of wash steps is feasible because of the ability of the flow cytometer to distinguish cell-bound and free fluorophore at free fluorophore concentrations of up to several hundred nM.

The specific assay described here exploits the high-content measurement capabilities of the flow cytometer to probe simultaneously two different receptors in a common assay volume – a duplex assay format. Cells expressing different receptors are color-coded to allow their distinction in a red-fluorescence detection channel during the analysis. The ligand-binding response is recorded in a separate green-fluorescence detection channel. An important advantage of the duplex approach, and of higher levels of assay multiplexing compatible with flow cytometry, is the ability to obtain both activity and selectivity information as an integral part of a primary screen. This assay is performed in 384-well plates, facilitated by the unique capability of HT flow cytometry with the HyperCyt® platform. In addition, this assay detects ligands to two related G-protein-coupled receptors (GPCR): the formylpeptide receptor (FPR) and formylpeptide receptor-like 1 (FPRL1) *(6)*. The FPR, a prototypic GPCR in the chemoattractant receptor superfamily, has been implicated in mediating inflammatory responses of clinical significance *(7, 8)*. FPRL1 is a related GPCR that shares 69% identity at the amino-acid level with FPR *(9, 10)*. Although even micromolar levels of potent FPR agonists such as the *N*-formylpeptide fMLF only weakly activate FPRL1 *(11)*, a number of host-derived FPRL1 peptide agonists have been identified in chronic inflammation-associated diseases such as systemic amyloidosis *(12)*, Alzheimer's disease *(13, 14)*, and prion diseases *(15, 16)*. This assay has been used at the New Mexico Molecular Libraries Screening Center to screen more than 25,000 compounds from the NIH Small Molecule Repository to identify selective ligands for each of the receptors (*see* NMMLSC assays at http://pubchem.ncbi.nlm.nih.gov/).

2. Materials

1. Tissue culture medium (TCM): RPMI-1640 medium (Mediatech, Herndon, VA) supplemented with 10% fetal bovine serum (heat-inactivated; US Biotechnologies, Parker Ford, PA), 2 mM l-glutamine, 10 U/mL penicillin, 10 μg/mL streptomycin (Mediatech), 10 mM HEPES (Mediatech), and

4 µg/mL CIPRO (Bayer Pharmaceuticals, Pittsburgh, PA). This medium may be stored at 4°C for up to 3 weeks. TCM is supplemented with 2.5 µg/mL amphotericin B (Mediatech) for culture of transfected RBL-2H3 cells.

2. A solution of 0.25% Trypsin-1 mM EDTA (Invitrogen, Carlsbad, CA) is used for passaging and preparation of RBL-2H3 cells.
3. *N*-formyl-methionine-leucine-phenylalanine-phenylalanine peptide (fMLFF) (Sigma, St. Louis, MO) is a peptide that selectively binds FPR.
4. Tryptophan-lysine-tyrosine-methionine-valine-D-methionine (Wpep) is a hexapeptide reported to bind with high affinity to both FPRL1 and FPR, but with a $K_d < 1$ nM for FPRL1 *(17, 18)* as compared with 95 nM for FPR *(17)*.
5. Fluorescein isothiocyanate-labeled Wpep (Wpep-FITC) (custom synthesis; Bachem, Torrance, CA) is used to quantify free FPR and FPRL1 receptors (*see* **Note 1**).
6. Peptide dilution buffer (PDB): 110 mM NaCl, 30 mM HEPES, 10 mM KCl, 1 mM $MgCl_2$, 10 mM glucose, and 0.1% bovine serum albumin. This solution is prepared fresh weekly and stored at 4°C.
7. Peptide solutions: peptides are stored as 1 mM or 100 µM aliquots in 100% DMSO (Omnisolv, EMD Chemicals, San Diego, CA) at –20°C. On each day of assay, aliquots are freshly thawed, diluted in PDB, and used immediately. The following solutions are adequate for preparation of four assay plates:
   a. Wpep-FITC: 4.5 µL of 100 µM Wpep-FITC in 15-mL cold PDB (15 nM Wpep-FITC).
   b. Blocking control solution: 20 µL of 100 µM Wpep, 3.6 µL of 1 mM fMLFF, 176-µL DMSO in 10-mL PDB (360 nM fMLFF, 200 nM Wpep, 2% v/v DMSO).
8. FuraRed staining solution: Add 46-µL DMSO to each of 2× 50-µg Fura Red™, AM (Invitrogen) for a 1 mM stock solution.
9. Cell strainer: a receptacle with a 70-µm nylon mesh filter (BD Biosciences, San Jose, CA) that is used to remove clumped cells from cell suspensions.
10. Compound dilution plate: a polypropylene 384-well plate with conical-bottom wells (Greiner, Monroe, NC) used for diluting compounds prior to addition to the assay sample.
11. Assay plate: a polystyrene 384-well plate with flat-bottom, small-volume wells (Greiner) in which ligand-binding assays are performed.
12. ThermalSeal sealing film adhesive plate-well covers (Excel Scientific, Wrightwood, CA) are used to cover assay plate wells during incubations.

13. Contrad70 (Fisher Scientific, Pittsburgh, PA) is diluted with water to 2.5% (v/v) for cleaning of microfluidics tubing.
14. Test compounds are obtained from a variety of sources and stored at –25°C in DMSO. Source 384-well plates for use in assays contain 1 mM compound and are sealed with Seal and Sample aluminum foil adhesive plate well covers (Beckman Coulter, Fullerton, CA).

3. Methods

3.1. Preparation of Cells

1. Cells are passaged twice weekly in 175-cm^2 plastic tissue culture flasks. Cell density should be no less than 6×10^5 cells/mL after the split. Cells expressing 100,000–250,000 receptors per cell are used in assays (*see* **Note 2**).
2. RBL/FPRL1, RBL-2H3 cells expressing human FPRL1, are grown as adherent cell cultures in TCM supplemented with 2.5 μg/mL amphotericin B. Cells may be grown beyond confluence prior to harvesting, as long as fresh TCM is added to maintain neutral pH. Use three flasks per four assay plates of RBL/FPRL1.
3. Rinse each flask with 25 mL of sterile PBS, add 10 mL 0.25% trypsin–EDTA, and incubate at 37°C until cells are detached.
4. Pool cells in 40-mL TCM, centrifuge 10 min at 450 × *g*, resuspend at 4×10^6/mL in PDB, and store on ice.
5. U937/FPR, U937 cells expressing human FPR, are grown as 100-mL suspensions in TCM and color-coded with FuraRed for use in assays. Use one flask of U937/FPR cells per four assay plates. Cell density should not exceed 1×10^6/mL at the time of harvest.
6. Centrifuge cells 10 min at 450 × *g*. Resuspend cells in PDB at 8×10^6 cells per 15 mL of TCM.
7. Add 90-μL FuraRed-AM solution per 15-mL cells and incubate 15 min at 37°C.
8. Wash loaded cells twice by centrifugation in PDB (15 mL per wash).
9. Resuspend the cell pellet in the RBL/FPRL1 cell suspension so that U937/FPR cells are also brought to a concentration of 4×10^6/mL. Store the cell mixture on ice.
10. Just prior to dispensing into the assay plate, the cell mixture is passed through a cell strainer to remove cell clumps.

3.2. Setup of 384-Well Dilution and Assay Plates

These instructions assume the use of automated pipetting robots and a liquid-dispensing instrument. A Biomek NX Multichannel pipetting robot (Beckman Coulter) equipped with a 384-P20 pod is used for plate-to-plate liquid transfers (**steps 3** and **4**). A Biomek NX Span-8 pipetting robot (Beckman Coulter) is used for addition of control reagents to plates (**steps 5** and **6**). A MAP-C2 liquid-dispensing instrument (TiterTek) is used for dispensing cells and reagents into microplate wells (**steps 2**, **7**, and **9**). Other comparable instruments capable of accurately pipetting 1- to 5-μL fluid volumes and dispensing 5-μL fluid volumes should also be suitable. Manual pipetting with multichannel pipettors is also feasible. However, it is not recommended for this assay configuration due to time constraints on some of the assay steps that will be difficult to satisfy.

1. Remove the 384-well plates containing test compounds from the freezer and allow them to warm to room temperature before removing the adhesive aluminum foil lid (*see* **Note 3**).
2. Dispense 49-μL ice-cold PDB to wells in columns 3–22 of a 384-well compound dilution plate.
3. Pipet 1 μL of test compounds from a compound storage plate to the matching, PDB-containing wells in columns 3–22 of the dilution plate.
4. Pipet 5 μL of test compounds from the dilution plate into the corresponding wells of a separate 384-well assay plate (columns 3–22). Use three aspirate/dispense pipetting cycles to mix the test compounds in the dilution plate wells before transfer.
5. Pipet 5-μL peptide blocking control solution into positive control wells (columns 2 and 23).
6. Pipet 5-μL PDB/2% DMSO (v/v) into negative-control wells (columns 1 and 24).
7. Into each well, dispense 5 μL of a cell mixture containing 50,000 of each cell type. The cell mixture contains U937/FPR (prelabeled with FuraRed) and RBL/FPRL1 cells, each at 4×10^6 cells/mL, and is chilled on ice prior to addition (*see* **Note 4**).
8. Seal the plate wells with ThermalSeal Film and incubate the assay plate at 4°C for 30 min, periodically inverting the plate to maintain cells in suspension (*see* **Note 5**).
9. Remove the sealing film and dispense 5-μL Wpep-FITC solution into each well (*see* **Note 4**).
10. Reseal the plate wells with ThermalSeal Film and incubate the plate at 4°C for 30 min or more while rotating.
11. Move the plates one at a time from the rotator to the deck of the HyperCyt® autosampler. Wait until sampling of one plate is completed before moving the next to the deck.

3.3. Setup of the HyperCyt® Platform and Flow Cytometer

These instructions assume the use of a HyperCyt® platform with a Dako CyAn flow cytometer. The HyperCyt® platform has been successfully used with several other flow cytometers, but as of this writing only the CyAn has been validated for uninterrupted analysis of 384-well plate assays. A critical flow cytometer feature is the availability of a high-resolution time parameter (data binning at intervals of 100 ms or less) that can be collected continuously for 10 or more min during data acquisition (*see* **Note 6**). The HyperCyt® platform must be properly configured and calibrated on a daily basis before use.

1. Align the HyperCyt® sampling probe to the four corner wells of a 384-well assay plate. The autosampler controlling software, HyperSip, has a programmed function to assist the user in this operation. First the probe is moved to Well A1 at which is defined the *X–Y* axis location corresponding to the center of the well and two Z-axis height locations corresponding to the bottom of the well and the height above the plate at which the probe will travel between wells. The probe is then moved to the other corner wells (A24, P24, and P1) in sequence to define their *X–Y* axis locations. Alignment once a day is typically sufficient and requires ~5 min.
2. If necessary, connect the HyperCyt® fluidics tubing to the sample uptake pathway of the flow cytometer (*see* **Note 6**). This requires <1 min.
3. Clamp the fluidics tubing in the peristaltic pump.
4. Move the sampling probe to the rinse station and turn on the peristaltic pump (15 rpm).
5. The sampling probe is repeatedly moved up and down, in and out of the rinse station solution (PDB), to generate a moving stream of discrete fluid volumes (~1 in in length) separated by air bubbles.
6. Adjust the peristaltic pump clamping pressure. When adjusted properly, there should be uniform air bubbles on both sides of the pump. If the bubbles are broken up on the flow cytometer side of the pump, the clamp has probably been tightened too much. Finding just the right tension will sometimes take a bit of time and patience (~5–10 min or less for the experienced operator). Once set, clamping pressure typically remains stable over the course of several days without requiring operator intervention.
7. Prior to the first sampling run, and between subsequent runs, the sampling probe is moved to the rinse station while the peristaltic pump continues to run. The 0.1% BSA in the PDB is an important surfactant that helps flush particles from the sample line. The presence of at least 0.1% BSA (or some other equivalent protein) in all solutions helps minimize particle carryover. It is also good practice to run detergent [2.5% Contrad-70 (v/v)] followed by distilled water through the fluidics tubing at the beginning and end of each day.

8. Use the *Plate Setup* feature of HyperSip software to create a worklist that controls the pattern with which wells will be sampled. Sample identifying information for each well may also be included in the worklist for postanalysis annotation of flow cytometry data. Worklists may be created in advance and automatically loaded just prior to assay plate sampling.
9. A preconfigured flow cytometry software protocol is loaded into the flow cytometer that defines the lasers and optical detectors to be used, the sensitivity settings for each detector, and graphical plots to be displayed during data acquisition. Quantitative analysis of the present assay requires acquisition of green and red fluorescence emission wavelength bands (FL1, 530 ± 20 and FL4, 630 ± 15 nm, respectively), two light-scatter signals (*FS* forward scatter and *SS* side scatter), and a high-resolution time parameter (data binned at intervals of 100 ms or less). All optical signals are excited by a single laser line of 488-nm wavelength.

3.4. Sampling and Data Acquisition

1. Position an assay plate on the HyperCyt® autosampler deck.
2. Initiate HyperCyt® sampling and flow cytometer data acquisition.
3. The sampling probe moves from well to well, sampling from each and moving to the next without an intervening probe wash step. The probe is programmed to remain in each well aspirating sample for 900 ms and in the air between wells aspirating air bubbles for 400 ms. The peristaltic pump that mediates liquid transfer runs continuously during sampling of the entire plate (*see* **Note 7**).
4. The flow cytometer continuously acquires data during sampling of the entire plate and for about 40 s thereafter. This allows time for the last sample to transit the length of the sampling tube from the plate to the point of analysis. During this interval, the sampling probe is at the rinse station moving up and down while the peristaltic pump continues to run (*see* **Note 8**).
5. At the end of data acquisition, the single data file is saved in flow cytometry standard (FCS) 3.0 file format. The total time required for sampling, data acquisition and data-file storage is typically 11–12 min.

4. Analysis of Flow Cytometry Data

1. The flow cytometry data file is loaded into IDLeQuery analysis software.
2. Four graphical plots of the data are created: (a) FL1 vs. FL4, (b) FS vs. SS, (c) FS vs. FL4, and (d) time vs. number of events.

3. The fluorescence compensation control is used to correct the green fluorescence signal of Wpep-FITC for the small fraction of cell-associated Fura Red fluorescence that is picked up by the FL1 detector. The FL1 vs. FL4 plot is used to evaluate compensation. Compensation is correct when the two cell populations are aligned one atop the other and the leftmost tail of the U937/FPR population distribution is extended to a similar extent as in the RBL/FPRL1 population (**Fig. 1**) (*see* **Note 9**).

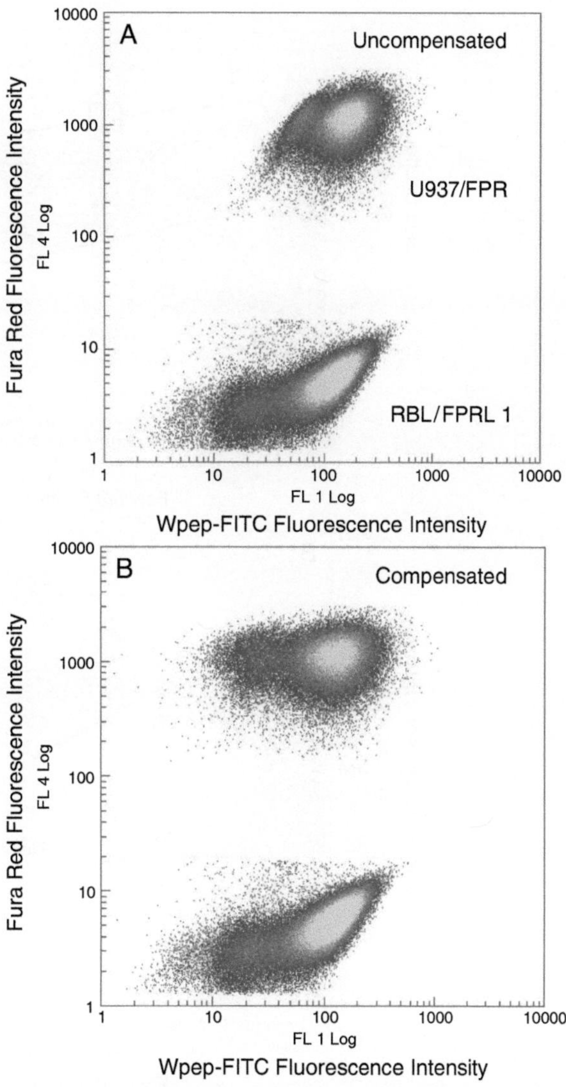

Fig. 1. Fluorescence compensation to correct for red fluorescence spillover into the green fluorescence detection channel. (**A**) Uncompensated fluorescence data. (**B**) Fluorescence profile after compensation. Approximately 3% of the signal detected in the red FL4 channel for each event was subtracted from the corresponding signal detected in the green FL1 channel. The plots represent combined data from all 384 wells.

4. In the FS vs. SS plot, an electronic gate (gate 1) is constructed to exclude dead cells and debris from the analysis (**Fig. 2a**).
5. In the FS vs. FL4 plot, two electronic gates are constructed, one about the bright red fluorescent U937/FPR cell population (gate 2) and the other about the dim red fluorescent RBL/FPRL1 cell population (gate 3) (**Fig. 2b**).
6. The time vs. number of events plot is displayed in the specialized Time Bins Analysis window (**Fig. 3**). Each peak in the

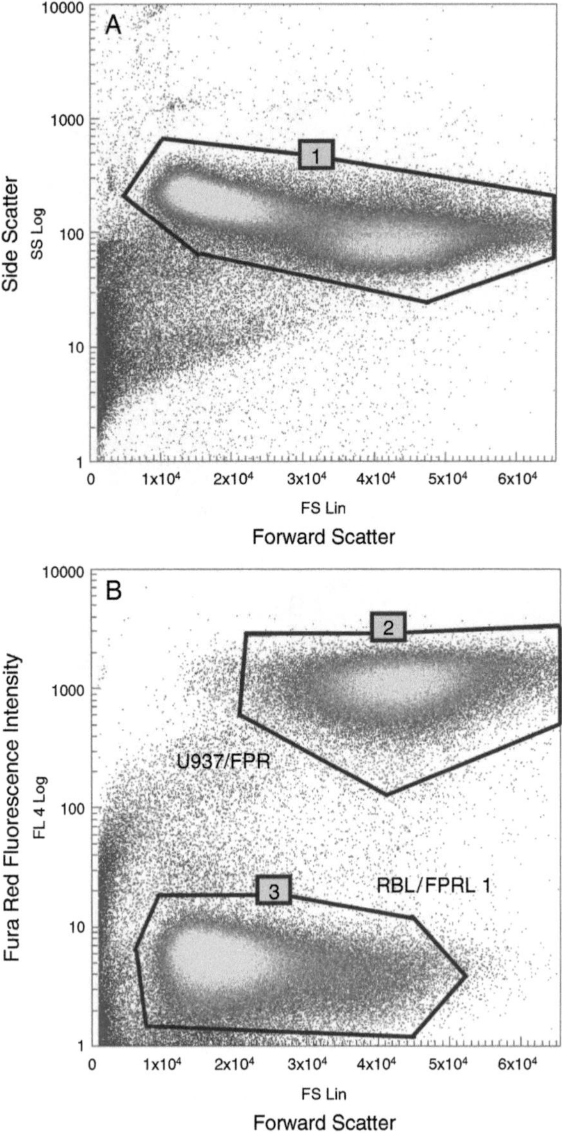

Fig. 2. Electronic gates required for analysis. (**A**) A light scatter gate (gate 1) is constructed to exclude debris and nonviable cells from the analysis. (**B**) Two additional gates are constructed based on the color-coding signals in the red fluorescence channel. One encloses red fluorescent U937/FPR cells (gate 2) and the other encloses unstained RBL/FPRL1 cells (gate 3). The plots represent combined data from all 384 wells.

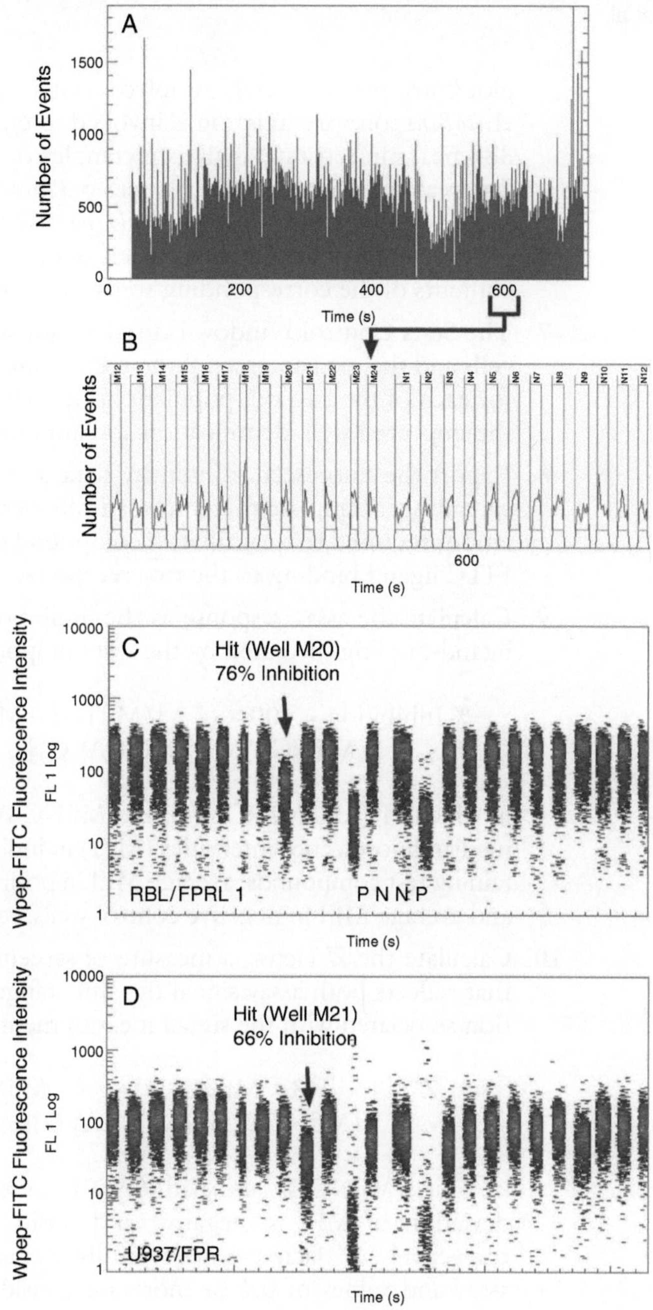

Fig. 3. Resolution and analysis of cells sampled from individual wells. (**A**) Display of data from all 384 wells in the Time Bins Analysis window in which the numbers of events detected at 100-ms time intervals are plotted as a function of time. (**B**) A magnified view of data from 25 wells measured in the vicinity of the 600-s time point [boundaries indicated below the time axis in panel (**A**)]. A software peak-detection algorithm identified 384 sets of data and enclosed each in a separate rectangular analysis region or Time Bin. Data sets were segregated on the basis of time gaps in the data stream produced by the passage of air bubbles. The first set is from well M12, and the last from well N12. (**C**) Green-fluorescence data from RBL/FPRL1 cells (gate 3) in wells M12 to N12. The test compound in well M20 (*arrow*) caused a 75% inhibition of Wpep-FITC binding relative to fluorescence intensities measured in control wells. It did not detectably affect Wpep-FITC binding to U937/FPR cells in the same well [*see* panel (**D**)]. Data from two positive-control wells (P) and two negative-control wells (N) are also indicated. (**D**) Green-fluorescence data from U937/FPR cells (gate 2) in wells M12 to N12. Well M21 contained a test compound that caused 66% inhibition of Wpep-FITC binding and no detectable effect on Wpep-FITC binding to RBL/FPRL1 cells in the same well [*see* panel (**C**)]

plot corresponds to cells sampled from a separate well. The *AutoBins* software function is invoked to enclose each of the 384 peaks in separate analysis rectangles or bins. The worklist created earlier with the HyperSip *Plate Setup* function is imported using the IDLeQuery *Import Well ID* function to provide appropriate labels for each bin and to annotate the contents of the corresponding wells (*see* **Note 10**).

7. The Stats Control Window is used to display the number of cells and the median green fluorescence intensities calculated for each of the two cell populations in each well, along with the imported well identification and annotation data.
8. Export the fluorescence intensity data to a Microsoft Excel spreadsheet template that automatically performs the calculations required to quantify test compound effects on Wpep-FITC ligand binding to the two receptors.
9. Calculate the assay response as the inhibition of fluorescent ligand-binding mediated by the test compound:

$$\%\ \text{Inhibition} = 100 \times \{1 - [(\text{MfiTest} - \text{MfiPCAVG})/(\text{MfiNCAVG} - \text{MfiPCAVG})]\},$$

in which MfiTest, MfiPCAVG, and MfiNCAVG represent the median fluorescence intensities (MFI) in individual wells containing test compounds, average MFI in positive control wells, and average MFI in negative control wells, respectively.

10. Calculate the Z' factor, a measure of screening assay quality that reflects both assay signal dynamic range and data variation associated with the signal measurements *(19)*:

$$Z' = 1 - \{[(3 \times \text{MfiPCSD}) + (3 \times \text{MfiNCSD})]/(\text{MfiNCAVG} - \text{MfiPCAVG})\},$$

in which MfiNCSD and MfiPCSD represent the standard deviation of MFI in negative and positive control wells, respectively. Z' factor values typically average ~0.65 for this assay and values of 0.4 or more are considered to indicate acceptable assay quality.

5. Notes

1. Addition of the fluorescein label to the lysine residue of Wpep alters the affinity to a K_d of 1.8 nM for FPRL1 and 1.2 nM for FPR, as determined in equilibrium-binding assays at 4°C (data not shown).

2. Receptor expression is determined by incubating cells with 10 nM Wpep-FITC for 30 min at 4°C in the presence and absence of 1 μM fMLFF + 1μM Wpep. Specific peptide binding is computed as the difference in cell-median fluorescence intensity (MFI) between unblocked and blocked (fMLFF/Wpep present) conditions and this is transformed to the estimated number of receptors (ligand-binding sites) per cell by comparison to an MFI standard curve generated with Quantum FITC MESF calibration beads (Bangs Labs, Inc., Fishers, IN). If receptor expression decreases or becomes heterogeneous, cells are stained with Wpep-FITC as above and sorted in the flow cytometer to purify and expand cells with high receptor expression (typically 1–5% of cells with brightest fluorescence).
3. The time required for the thawing of compound solutions will depend upon the volume in each well and well geometry. Remove plate covers only after any condensation has evaporated, and minimize any subsequent exposure of compounds dissolved in DMSO to humidity. This is especially important if the compounds will be subjected to multiple freeze–thaw cycles. DMSO rapidly absorbs water and compounds will be increasingly prone to becoming insoluble during a freeze–thaw cycle as DMSO water content increases. Once thawed, the plates are centrifuged at 500 × *g* for 2 min at room temperature. This positions compound solutions uniformly in the bottom of the wells to facilitate accurate automated pipetting.
4. Cell- or reagent-dispensing must be done quickly in this step (<3 min) to minimize warming and cell settling.
5. Plates are placed on a custom-built system that rotates at 4 rpm to periodically invert the plates. The assay plate well geometry favors fluid retention in wells of an inverted plate due to surface tension. Alternatively, incubation can be performed under static conditions without continuous rotation of the plate. A pipet-mixing step to resuspend cells should be added at the end of the incubation if this option is selected.
6. At this writing, the HyperCyt® platform is available for purchase under a beta-testing agreement at IntelliCyt (http://www.intellicyt.com). A simple modification of the sample uptake fluidics pathway of the CyAn flow cytometer is required to allow it to interface with the HyperCyt® platform. One end of the OEM silicone sample transport tubing is detached from the bottom of flow cytometer sample interrogation cuvette. An 8-inch length of PVC tubing is attached there instead. A 1-inch length of 25G stainless steel tubing is used to join the free end of this PVC tubing to either i) the free end of the OEM silicone tubing to enable manual sample introduction, or ii) the free end of the PVC sample-transport tubing of

the HyperCyt® platform to enable HT sample introduction. Switching between the two modes of operation requires less than 1 min.

7. The peristaltic pump is rotated at 15 rpm to result in a sample flow rate of ~2 μL/s. Faster or slower rates are typically suboptimal, and can also result in increased particle carryover.
8. This generates a trailing train of bubble-separated liquid volumes to ensure that the last samples are pushed through the data-analysis point under the same conditions as the others. Air bubbles are compressible, water is not. A trailing train of PDB would cause the last of the sample-separating air bubbles to compress more than the preceding, pushing the last samples through more rapidly.
9. The leftmost tail represents cells with low green fluorescence intensity from wells containing compounds that block Wpep-FITC binding. This includes the 32 positive-control wells and wells containing active test compounds.
10. Several redundant sources of information are used to ensure accurate indexing of time-resolved flow cytometry data to source samples. Current practice is to index microplate wells at fixed intervals using the green FL1 fluorescence-intensity profiles of the control wells. A pattern is generated by the positive-control wells that distinguish the boundaries of each 24-well span (**Fig. 4**). Temporal gaps in the data caused by air bubbles provide additional reference signals for indexing data from individual wells. This combined indexing approach has proven to be quite reliable, but not without occasional ambiguities of automated data analysis, requiring human intervention to resolve. Additional overlapping and redundant indexing signals will likely extend system-performance capabilities for full automation.

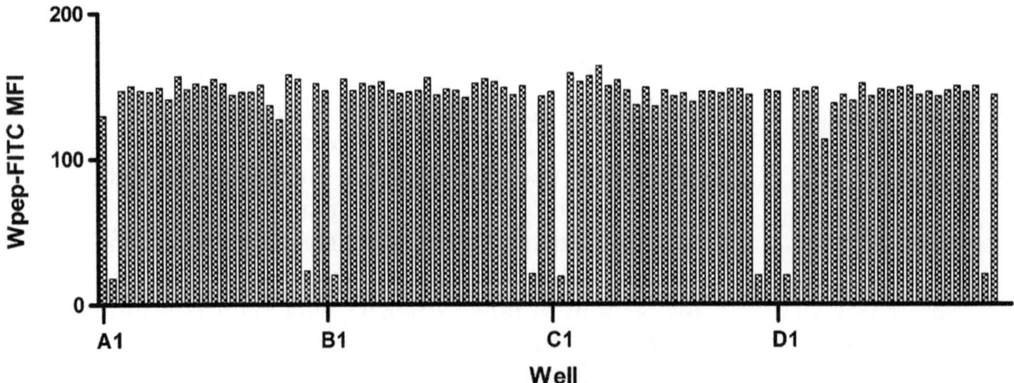

Fig. 4. Repetitive fluorescence intensity pattern provides supplementary sample indexing information. Illustrated are green fluorescence data from the first 96 wells of a 384-well plate. Wells are sampled from left to right (columns 1–24) one row at a time, starting with the top row (row A) and proceeding downward. This results in a repetitive pattern of low median fluorescence intensity (MFI) signals marking the 2nd and 23rd wells of each row, the locations of the positive control samples. This information facilitates identification of a row in which an indexing error has occurred.

Acknowledgments

This work was supported by NIH R03 MH076381-01, U54 MH074425-01, the New Mexico Molecular Libraries Screening Center, the University of New Mexico Shared Flow Cytometry Resource, and Cancer Research and Treatment Center.

References

1. Kuckuck, F. W., Edwards, B. S., and Sklar, L. A. (2001) High throughput flow cytometry. *Cytometry* 44, 83–90.
2. Ramirez, S., Aiken, C. T., Andrzejewski, B., Sklar, L. A., and Edwards, B. S. (2003) High-throughput flow cytometry: validation in microvolume bioassays. *Cytometry A* 53, 55–65.
3. Young, S. M., Bologa, C., Prossnitz, E., Oprea, T. I., Sklar, L. A., and Edwards, B. S. (2005) High-throughput screening with HyperCyt flow cytometry to detect small molecule formylpeptide receptor ligands. *J. Biomol. Screen.* 10, 374–382.
4. Edwards, B. S., Bologa, C., Young, S. M., Balakin, K. V., Prossnitz, E., Savchuck, N. P., Sklar, L. A., and Oprea, T. I. (2005) Integration of virtual screening with high-throughput flow cytometry to identify novel small molecule formylpeptide receptor antagonists. *Mol. Pharmacol.* 68, 1301–1310.
5. Edwards, B. S., Young, S. M., Oprea, T. I., Bologa, C., Prossnitz, E., and Sklar, L. A. (2006) Biomolecular screening of formylpeptide receptor ligands with a sensitive, quantitative, high-throughput flow cytometry platform. *Nat. Protocols* 1, 59–66.
6. Migeotte, I., Communi, D., and Parmentier, M. (2006) Formyl peptide receptors: a promiscuous subfamily of G protein-coupled receptors controlling immune responses. *Cytokine Growth Factor Rev.* 17, 501–519.
7. Le, Y., Murphy, P. M., and Wang, J. M. (2002) Formyl-peptide receptors revisited. *Trends Immunol.* 23, 541–548.
8. Oppenheim, J. J., Zachariae, C. O., Mukaida, N., and Matsushima, K. (1991) Properties of the novel proinflammatory supergene "intercrine" cytokine family. *Annu. Rev. Immunol.* 9, 617–648.
9. Murphy, P. M. (1996), in Chemoattractant Ligands and Their Receptors (Horuk, R., ed.), CRC Press, Boca Raton, FL, pp. 269–299.
10. Prossnitz, E. R., and Ye, R. D. (1997) The *N*-formyl peptide receptor: a model for the study of chemoattractant receptor structure and function. *Pharmacol. Ther.* 74, 73–102.
11. Ye, R. D., Cavanagh, S. L., Quehenberger, O., Prossnitz, E. R., and Cochrane, C. G. (1992) Isolation of a cDNA that encodes a novel granulocyte *N*-formyl peptide receptor. *Biochem. Biophys. Res. Commun.* 184, 582–589.
12. Snipe, J. D. (1990), in Immunophysiology: The Role of Cells and Cytokines in Immunity and Inflammation (Oppenheim, J. J. and Shevac, E. M., eds.), Oxford University Press, New York, pp. 259–273.
13. Lambert, M. P., Barlow, A. K., Chromy, B. A., Edwards, C., Freed, R., Liosatos, M., Morgan, T. E., Rozovsky, I., Trommer, B., Viola, K. L., Wals, P., Zhang, C., Finch, C. E., Krafft, G. A., and Klein, W. L. (1998) Diffusible, nonfibrillar ligands derived from Abeta1-42 are potent central nervous system neurotoxins. *Proc. Natl. Acad. Sci. USA* 95, 6448–6453.
14. Kalaria, R. N. (1999) Microglia and Alzheimer's disease. *Curr. Opin. Hematol.* 6, 15–24.
15. Brown, D. R., Schmidt, B., and Kretzschmar, H. A. (1996) Role of microglia and host prion protein in neurotoxicity of a prion protein fragment. *Nature* 380, 345–347.
16. Le, Y., Yazawa, H., Gong, W., Yu, Z., Ferrans, V. J., Murphy, P. M., and Wang, J. M. (2001) The neurotoxic prion peptide fragment PrP(106-126) is a chemotactic agonist for the G protein-coupled receptor formyl peptide receptor-like 1. *J. Immunol.* 166, 1448–1451.
17. Bae, Y. S., Song, J. Y., Kim, Y., He, R., Ye, R. D., Kwak, J. Y., Suh, P. G., and Ryu, S. H. (2003) Differential activation of formyl peptide receptor signaling by peptide ligands. *Mol. Pharmacol.* 64, 841–847.
18. Bae, Y. S., Yi, H. J., Lee, H. Y., Jo, E. J., Kim, J. I., Lee, T. G., Ye, R. D., Kwak, J. Y., and Ryu, S. H. (2003) Differential activation of formyl peptide receptor-like 1 by peptide ligands. *J. Immunol.* 171, 6807–6813.
19. Zhang, J. H., Chung, T. D., and Oldenburg, K. R. (1999) A simple statistical parameter for use in evaluation and validation of high throughput screening assays. *J. Biomol. Screen.* 4, 67–73.

Chapter 12

High-Throughput Real-Time PCR for Detection of Gene-Expression Levels

Bridget K. Wagner and Zoltan Arany

Summary

While many high-throughput screening campaigns involve the measurement of protein levels or locations, at times it is desirable to measure the levels of gene expression in response to small molecules. Here, we describe a method for capturing mRNA in multiwell plates following compound treatment and measuring gene expression using real-time PCR. This streamlined protocol provides complementary information to conventional phenotypic cell-based assays, and is especially useful in cases where the gene of interest is thought to serve a regulatory function in downstream cellular phenotypes.

Key words: High-throughput gene expression, mRNA capture, Real-time PCR, Reverse transcription.

1. Introduction

High-throughput screening (HTS) involves the rapid assessment of large numbers of compounds for desired biological effects. Advances in technology and affordability have made feasible the use of HTS in an academic setting. Methods have been developed to measure, for example, protein expression *(1)*, protein binding *(2)*, and protein localization *(3)*. More recently, measurements of nucleic acids in response to small molecules have been developed. The evaluation of gene-expression levels is often used *after* HTS in order to identify the targets of small molecules of interest. However, there are circumstances in which the levels of gene expression themselves are the most relevant measurement for screening. In cases such as these, one can imagine two

P.A. Clemons et al. (eds.), *Cell-Based Assays for High-Throughput Screening, Methods in Molecular Biology, Vol. 486*

DOI: 10.1007/978-1-60327-545-3_1

modalities for screening: first, the analysis of individual target-gene expression, which is especially appropriate for genes that are thought to serve regulatory functions, and second, the use of a multigene signature of a cellular phenotype, which works well for difficult-to-measure cellular outcomes. For the latter approach, the recent development of gene expression-based HTS (GE-HTS) *(4–6)* has enabled the simultaneous measurement of many transcripts, often in the context of identifying compounds that can induce cell-state switching, such as differentiation. This chapter will focus on the former screening paradigm, in which the analysis of individual target-gene expression is assessed in a high-throughput manner.

Overall, this method involves capturing mRNA from cells treated with compounds, and measuring gene expression using reverse transcription and real-time PCR (*see* **Note 1**). Cell lysis is performed within the compound-treated wells, and lysate is then transferred to a commercially available multiwell plate coated with oligo-dT for capture of mRNA. In-plate reverse transcription yields cDNA, which is in turn transferred to a PCR plate, where the gene of interest is amplified. Parallel measurement of a control gene allows normalization for mRNA content; this is particularly important for chemical screening, as compounds may nonspecifically affect cell number, viability, or per-cell mRNA content. This method results in the ability to assess the effects of a compound collection on individual gene expression.

2. Materials

2.1. Cell Culture

2.1.1. Plates

The choice of multiwell plates in which cells are cultured for screening is open to the discretion of the screener. In contrast to fluorescent and luminescence-based screening, the final step of detection will not be performed in the plates that are used for cell culture, so considerations such as clear bottoms or plate color are less important than in such cases. We have had success using Costar-brand clear cell-culture plates (Corning, Lowell, MA).

2.1.2. Culture Media

The protocol for capturing polyadenylated mRNA in multiwell plates can be used on adherent cells, suspension cells, or tissues. We have performed this protocol on adherent cells.

1. Dulbecco's modified Eagle medium (Invitrogen, Carlsbad, CA).
2. Fetal bovine serum (FBS) to a final concentration of 5–10% (Invitrogen).
3. Penicillin-streptomycin solution (Invitrogen).

4. Phosphate buffered saline (PBS), pH 7.4.
5. Trypsin–EDTA solution (0.25%) (Invitrogen).

2.2. Capture of Polyadenylated mRNA

1. TurboCapture 384 mRNA kit (Qiagen, Valencia, CA). This kit includes the following materials: 384-well plates, the wells of which contain immobilized oligo-dT; buffer for in-well cell lysis; and wash buffer.
2. β-mercaptoethanol (commercial 14.3 M solution).

2.3. Amplification and Detection

1. iScript cDNA synthesis kit (BioRad, Hercules, CA).
2. Thermal cycler, such as the DNA Engine Tetrad (GMI, Ramsey, MN), capable of holding multiwell plates.
3. RNase-free water.
4. At least two high-throughput optical PCR plates (ABI, Foster City, CA), one or more for the gene(s) of interest, and one for a gene-expression control gene. This number of plates is needed for each compound stock plate; if screening multiple compound plates in replicate, the need for PCR plates obviously increases in a geometric fashion.
5. SYBR-green containing 2× PCR master mix (ABI).
6. Optical plate seals (ABI).
7. ABI Prism 7900, or similar real-time PCR intrument, using SYBR green as detector.

2.4. Robotics

2.4.1. Liquid Dispensing

1. In the absence of suitable liquid-handling robotics, a common laboratory multichannel pipettor will suffice for much of this protocol. However, the workload involved in screening will increase dramatically.
2. We use a μFill (BioTek, Winooski, VT) liquid dispenser for washing cells with PBS, as it is possible to adjust the dispense rate to levels slow enough to prevent disruption of the cell monolayers in each well.
3. For steps involving more precious reagents, we use a Multidrop Combi (Thermo Electron, Waltham, MA), favoring this instrument due to its very low dead volume and ability to recapture remaining reagent following dispensing.
4. The μFill dead volume is ~13 mL, while that of the Multidrop equipped with the low-volume manifold is only 1.5–2 mL.

2.4.2. Liquid Aspiration

1. As is the case with liquid dispensing, a multichannel aspirating wand (V & P Scientific, San Diego, CA) could suffice to remove media from multiwell plates.
2. However, we use an ELx405 plate washer (BioTek, Winooski, VT) to remove media in an automated fashion. This instrument

is capable of aspirating 96 wells in one motion, with Cartesian coordinates specified by the user.

2.4.3. Liquid Transfer

1. There are several steps in which the entire contents of a 384-well plate must be transferred to a new plate. While it may be feasible to imagine the use of a multichannel pipettor, there exist commercially available liquid-transfer robotics systems.
2. We use the CyBi-Well Vario (CyBio, Boston, MA), which can be equipped with tip trays containing 96 or 384 pipet tips, capable of transferring up to 250 or 25 μL each, respectively.

3. Methods

3.1. Cell Culture

1. Seed cells in multiwell plates. Again, the choice of plates used for this step is open to the user. To ensure that we had sufficient mRNA for detection, we used Costar brand 96-well plates containing a confluent monolayer of primary mouse muscle cells.
2. Thus, four 96-well plates of cells are needed for each 384-well compound plate.
3. In general, 10,000 cells per well of a 96-well plate results in a confluent layer; for mouse models of muscle that require differentiation, such as C2C12, we seed 8,000 per well and start differentiation the following day by replacing with media containing 2% horse serum (*see* **Note 2**).
4. Such cells require 4–6 days for differentiation, but the primary muscle cells, which have been isolated and frozen in liquid nitrogen, express markers of differentiation in 1 day.
5. Allow cells to adhere to wells. While many cell lines will adhere to the cell culture-treated multiwell plate in several hours, we allow the cells to incubate at least overnight in a water-jacketed cell-culture incubator containing 5% CO_2 and at 95% humidity.

3.2. Compound Treatment

1. Replace plates with fresh media. The pH and nutrient levels of the cell-culture media do not remain stable in multiwell plates for more than roughly 2 or 3 days, so we typically replace the plates with 100 μL/well fresh media immediately before adding compounds.
2. Add compounds. A more detailed discussion of compound sources is addressed elsewhere *(7, 8)*. There exist several technologies for compound treatment, including pin-transfer *(9)*, acoustic transfer *(10)*, or small-molecule microarrays *(11)*.

3. Using the first method listed in **step 2**, we used a CyBi-Well (CyBio) equipped with a 96-pin array to add 100-nL compound to each well of the 96-well plate of cells. The steel pins are dipped into a compound plate, and subsequently dipped into the cell-culture media, allowing the dispersal of compound (*see* **Note 3**).
4. Washing the pin array in DMSO (the solvent in which the compounds are dissolved) and methanol allows this process to be repeated.
5. This process results in a 1,000-fold dilution of compound, typically yielding screening concentrations of 10–20 μM.
6. We incubated the cells with compounds for 24 h, though the amount of treatment time depends on the immediacy of induction expected for the gene of interest.
7. To maximize gas exchange between stacks of multiwell plates, we used specially designed plate carriers containing shelf-like slots for each plate (please contact the corresponding author for specifications).

3.3. Preparing mRNA

1. Wash cells once with PBS, using either 100 μL/well (for a 96-well plate) or 50 μL/well (for a 384-well plate). When using a liquid-dispensing robot, be sure to employ a low dispense rate to prevent cells from becoming dislodged from the bottom of the well.
2. Aspirate wells fully, ensuring that no liquid remains in the well. Manual aspiration works well for this step, as do several other techniques (*see* **Note 4**).
3. Add β-mercaptoethanol to the commercial lysis buffer to a final concentration of 1% (v/v), and add 50-μL lysis buffer to each well.
4. Incubate plates for 5 min at room temperature. No shaking is required at this step (*see* **Note 5**).
5. Transfer 30 μL of the lysate to the TurboCapture plate. As mentioned earlier, it is technically possible to use a multichannel pipettor for this transfer step. However, an automated liquid handler speeds up the process considerably (*see* **Note 6**). We used a CyBi-Well Vario (CyBio) equipped with disposable 96-tip arrays, thus requiring four tip trays for each 384-well TurboCapture plate. We have found that transferring 30 μL of the 50-μL lysate yields sufficient transcript detection, and prevents any problems resulting from the pipet tips hitting the bottom of the cell-culture plate.
6. Incubate for 60–90 min with shaking at 100 rpm (*see* **Notes 7** and **8**). Any flat orbital shaker will suffice.
7. Wash plates three times with the proprietary wash buffer included with the TurboCapture kit, making sure that the last step is aspirated fully (*see* **Note 4**).

3.4. RT-PCR

1. To the now-empty TurboCapture plate, add 5 μL/well iScript cDNA synthesis mix. The use of random primers (rather than oligo-dT) is favored here, since the mRNA poly(A) tails are already annealed to the oligo-dT on the plates. The use of no primers is not a good option, as it will result in cDNA that is covalently bound to the plate.
2. Program thermal cycler, and allow reverse transcription to proceed. We used the scheme recommended by the manufacturer: 5 min at 25°C, 30 min at 42°C, and 5 min at 85°C (*see* **Note 9**).
3. Add 20 μL/well RNase-free water, using a liquid dispenser such as those listed in "Materials." This step results in 25-μL total cDNA solution (*see* **Note 10**).
4. For each TurboCapture plate, prepare two PCR plates: one for your transcript of interest, and one for the control transcript (e.g., TATA box-binding protein, or TBP). To each well, add 3-μL master mix (2.5 μL of 2× master mix, plus 0.25 μL of 1 μM each DNA primer to be used for PCR amplification). We have used standard two-probe quantitative PCR, with free SYBR in the mix, with good outcomes. In principle, TaqMan probes could be used at this point.
5. Add 2 μL/well cDNA to each PCR plate. The CyBi-Well Vario, equipped with a 384-pipet tip tray, can attain this precision, and we found this instrument very well suited for this step, as cDNA could be pipetted iteratively into multiple PCR plates, depending on the number of transcripts to be evaluated (*see* **Note 11**).
6. Perform PCR, using an automated instrument such as listed in "Materials." We have used the standard ΔΔCt method for calculating gene expression (*see* **Note 12**), roughly as follows. An arbitrary threshold is chosen, and the cycle number at which signal for the gene of interest crosses that threshold in each well is recorded. This value is then subtracted from the value acquired when measuring the TBP control. That value, in turn, is subtracted from the average value similarly acquired from control wells (typically a few dozen). The output represents the log (base 2) of fold-induction of gene expression.

4. Notes

1. Before commencing with full-scale small-molecule screening, we suggest careful assay development. For example, one should determine the coefficient of variation (CV) within the data. This can be determined by dividing the arithmetic mean

of the data by the standard deviation – in this case, normalized cycle number. We measured gene expression on an entire 384-well plate, half of which was treated with 0.1% DMSO, corresponding to the final concentration of solvent after compounds are pin-transferred into these plates, and half of which contained a positive control for increased expression of the gene of interest.

2. We have tested this protocol with other cell types, and noted that primary mouse adipocytes, upon lysis, result in a mixture that is too viscous for reliable pipetting using our liquid transfer systems. We suggest performing a small-scale test lysis on the cells of interest before proceeding with large-scale assay development.
3. During the course of any HTS campaign, it is important to keep compound and cell-culture plates in the same orientation, so as to ensure that well A01, for example, truly receives the compound residing in well A01 of the compound plate. In this protocol, we have described several steps requiring transfer of materials: compound addition to the cell-culture plate, consolidation from 96-well format to 384-well format, and transfer of cDNA from TurboCapture to PCR plates. Thus, it is vitally important to maintain orientation uniformity throughout the protocol, either by careful manual labeling or by the use of, for example, barcoding and scanning techniques.
4. There are several steps in this protocol that call for "full aspiration" of the liquid in each well. We have observed that automated plate washers tend to leave a very small but nonzero amount of liquid in each well (1–3 μL/well). A few simple methods are effective at removing this last volume. First, the plate may be centrifuged upside-down. In this case, the plate is placed in a centrifuge plate carrier, on top of several paper towels, and spun for 1 min at 1,000 rpm. This method is the surest way of removing all liquid from a plate, and is sufficiently gentle that disruption of the immobilized mRNA is prevented. Second, and a perhaps even more low-technology option, is that one may literally shake the remaining liquid volume out of the plate; this involves turning the plate upside down over a biohazard bag and applying a rapid downward snap.
5. The addition of lysis buffer to the cell-culture plates will cause lysis to occur nearly instantaneously. The user can verify this lysis by microscopic examination. There is no need for shaking. The 5-min incubation is indicated in order to ensure full lysis, and appears to be a conservative estimate.
6. When deciding between using a multichannel pipettor and liquid-dispensing robotics, it is useful to consider the number of plates to be processed, and the dead volume required. For example, precious reagents such as the PCR mix, at 3 μL/well,

require 1.15 mL per 384-well plate. The dead volume for the Multidrop Combi is 1.5–2 mL, more than 100% of what is needed for a single plate. However, if ten plates are being prepared at a sitting, 11.5-mL PCR mix is required, and the dead volume is now 13–18% of the total volume, making the automated dispensing more attractive and economical.

7. Once lysate is added to the TurboCapture plates, they should be centrifuged to assure that the lysate is in contact with the bottom of the well. One minute at 1,000 rpm is sufficient for this purpose.
8. The TurboCapture plate containing cell lysate should be incubated with shaking for at least 60 min at room temperature, but we have not observed better assay results by increasing that length of time.
9. For the reverse transcription step, a multiplate thermal cycler is recommended, as there is a 40-min interval for each plate. Thus, the ability to parallelize this process will result in the ability to process more plates per day.
10. Following the addition of RNase-free water, there is no need to transfer the cDNA to a new plate before distributing aliquots to PCR plates. The reduction in temperature from 85°C to room temperature will *not* cause the cDNA to reanneal to the TurboCapture well surface.
11. If the choice is made to use a multichannel pipettor when distributing aliquots of cDNA, it is important to remember that the tips must be changed for every change of row or column. Otherwise, well-to-well contamination will occur. The use of an automated liquid-transfer robot is ideal, but this is an important consideration if using more manual methods.
12. We have found the reproducibility of this assay to be very high. Under our conditions, the standard deviation for the ΔΔCt value was less than 0.5 (i.e., 1.4-fold induction). We have found that performing PCR once using the control gene, such as *TBP*, is sufficient for subsequent PCR runs. Thus, with 25-μL total cDNA yield, and at 2-μL cDNA per PCR experiment, one could theoretically measure the expression levels of 12 genes, including the control gene. In reality, pipetting error and cost considerations result in this number being typically lower.

Acknowledgments

We are indebted to the screening staff at the Broad Institute, particularly Stephanie Norton and Jason Burbank, for their help with compound pin-transfer, calibrating and programming robotic equipment, and general advice. We thank Yanhong Ma, Tamara Gilbert, and Daniel Fass for technical advice and expertise. This work has been funded by NIH grant 1R21NS059440 (to Z.A.), and in part with Federal funds from the National Cancer Institute's Initiative for Chemical Genetics, National Institutes of Health, under Contract No. N01-CO-12400. The content of this publication does not necessarily reflect the views or policies of the Department of Health and Human Service, nor does mention of trade names, commercial products, or organizations imply endorsement by the US Government.

References

1. Stockwell, B. R., Haggarty, S. J., and Schreiber, S. L. (1999) High-throughput screening of small molecules in miniaturized mammalian cell-based assays involving post-translational modifications. *Chem. Biol.* 6, 71–83.
2. Koehler, A. N., Shamji, A. F., and Schreiber, S. L. (2003) Discovery of an inhibitor of a transcription factor using small molecule microarrays and diversity-oriented synthesis. *J. Am. Chem. Soc.* 125, 8420–8421.
3. Perlman, Z. E., Slack, M. D., Feng, Y., Mitchison, T. J., Wu, L. F., and Altschuler, S. J. (2004) Multidimensional drug profiling by automated microscopy. *Science* 306, 1194–1198.
4. Hieronymus, H., Lamb, J., Ross, K. N., Peng, X. P., Clement, C., Rodina, A., Nieto, M., Du, J., Stegmaier, K., Raj, S. M., Maloney, K. N., Clardy, J., Hahn, W. C., Chiosis, G., and Golub, T. R. (2006) Gene expression signature-based chemical genomic prediction identifies a novel class of HSP90 pathway modulators. *Cancer Cell* 10, 321–330.
5. Wei, G., Twomey, D., Lamb, J., Schlis, K., Agarwal, J., Stam, R. W., Opferman, J. T., Sallan, S. E., den Boer, M. L., Pieters, R., Golub, T. R., and Armstrong, S. A. (2006) Gene expression-based chemical genomics identifies rapamycin as a modulators of MCL1 and glucocorticoid resistance. *Cancer Cell* 10, 331–342.
6. Stegmaier, K., Ross, K. N., Colavito, S. A., O'Malley, S., Stockwell, B. R., and Golub, T. R. (2004) Gene expression-based high-throughput screening (GE-HTS) and application to leukemia differentiation. *Nat. Genet.* 36, 257–263.
7. Schreiber, S. L. (2000) Target-oriented and diversity-oriented organic synthesis in drug discovery. *Science* 87, 1964–1969.
8. Schreiber, S. L. (2005) Small molecules: the missing link in the central dogma. *Nat. Chem. Biol.* 1, 64–66.
9. Cleveland, P. H. and Koutz, P. J. (2005) Nanoliter dispensing for uHTS using pin tools. *Assay Drug Dev. Technol.* 3, 213–225.
10. Olechno, J., Ellson, R., Browning, B., Stearns, R., Mutz, M., Travis, M., Oureshi, S., and Shieh, J. (2005) Acoustic auditing as a real-time, non-invasive quality control process for both source and assay plates. *Assay Drug Dev. Technol.* 3, 425–437.
11. Bradner, J. E., McPherson, O. M., and Koehler, A. N. (2006) A method for the covalent capture and screening of diverse small molecules in a microarray format. *Nat. Protoc.* 1, 2344–2352.

Chapter 13

Interpretation of Uniform-Well Readouts

Serene Josiah

Summary

High-throughput screening (HTS) covers a range of measurements, from primary screens of either large libraries (>250 K) or small, focused collections (100–1,000 s) of test compounds, to secondary screens used to characterize the mechanism of action of a relatively small number of compounds. Data analysis of assay results from HTS relies upon assay performance and the control wells used to define the assay system. This chapter discusses parameters that must be considered when defining controls and plate maps for primary and secondary assays in HTS. Control wells and plate maps are suggested, which can generally be applied toward a variety of biochemical and cellular assays. The controls and plate-map options can be matched to the scale of the screening campaign; examples are primary screens with % inhibition or % activation as endpoints or secondary screens with IC_{50} or EC_{50} values as endpoints.

Key words: Background signal, Focused screens, High-throughput screening, Plate map, Secondary screening, Total signal, *Z′* factor.

1. Introduction

Control wells mapped to defined locations on assay plates (plate map) provide key information for HTS data analysis. Data obtained from control wells determine the utility of the assay data set and define the dynamic range of the assay. Data from control wells are also used to identify the variability within a screening assay, thus enabling the user to retain or discard well data for analysis. Without appropriate controls, the user runs the risk of overinterpreting screening results and pursuing molecules for further characterization that are really within the background noise of the screen. Alternatively, true positives will more likely be missed, particularly in cell-based assays, where the dynamic

P.A. Clemons et al. (eds.), *Cell-Based Assays for High-Throughput Screening, Methods in Molecular Biology, Vol. 486*

DOI: 10.1007/978-1-60327-545-3_1

range of the assay may be limited or difficult to establish. Control wells are defined by components of the HTS assay and detection technology. Plate maps provide the user with a reproducible configuration on a given plate of two well types: control wells and test wells. Plate maps for HTS are designed to maximize test wells, while providing statistically significant replicates of control wells for data-quality control. Plate maps are an integral part of the standardization of HTS automation and data management. This chapter discusses options for choosing and applying control wells and plate maps to HTS for primary and secondary screens.

High-throughput screening is a multidisciplinary process that brings several scientific fields together *(1, 2)*. There are several activities that must be integrated to ensure a successful HTS campaign. These activities can be generally subdivided into four categories: (1) development of an HTS assay, typically biochemical or cellular *(3, 4)*; (2) handling of unknown samples of chemical, peptide, protein, RNA, or DNA *(5, 6)*; (3) implementation of assay detection technology platform utilizing full robotics, workstation(s), or manual processes *(7)*; and (4) data management and analysis *(8, 9)*.

To integrate the aforementioned activities, a viable HTS campaign will be able to intercalate into the screen the appropriate controls and establish suitable throughput to enable the generation of a large dataset on a collection of test samples. Two key factors toward integrating the various activities of HTS and thus enabling data analysis are establishing screening controls, and plate maps. In this chapter, examples are tailored to small-molecule (SM) HTS. The plate format used in examples is the 96-well format since this format can be easily scaled up (384- and 1536-well, and plate-free conditions) for HTS applications or scaled down (for 6-well, 24-well) to accommodate challenging secondary screening applications.

1.1. Overview of the Four Categories of a HTS Campaign

A well-developed assay is a key component in obtaining useful data from an HTS campaign *(10)*. Assays tend to be either biochemical or cellular endpoints. Whole animal technologies are available (*C. elegans*, zebrafish, etc.) and are usually implemented by specialized organizations *(11, 12)*. Various assay technologies are available to the user, providing choices for assay sensitivity (wide dynamic range of assay window) and throughput (homogenous vs. heterogenous assays).

There are many points to consider for small-molecule screening during assay development prior to HTS and data analysis. Briefly, key points for biochemical assays impacting HTS are the stabilities of assay proteins at time points used for the reaction and detection events of the assay. For example, early knowledge of protein stability will set parameters for the number of plates and samples that can be tested at a given time. The time course of a biochemical reaction may be relatively short (i.e., period for initial rate of kinase activity may be in minutes), which will also influence the number of plates and samples that can be tested at a given time. Some key points for cellular assays are establishing the cell-line passage number that is acceptable for biological activity and the cell density necessary to measure biological activity. Providing stringent criteria for assays such as those mentioned earlier improve assay reproducibility and decrease the identification of false positives.

Options for assay controls include (a) positive controls, (b) total-signal controls, and (c) background-signal controls. The minimal controls needed to perform data analysis of HTS are the total-signal and background-signal controls. These two parameters provide information regarding the dynamic range of the assay and allow the user to quantify changes within the dynamic range (positive or negative from a reference point). A positive-control well with a known active agent in the assay (small molecule, protein, chelaters, etc.) at an IC_{50} or EC_{50} dose can also be useful for monitoring the dynamic range of the assay. Background-signal and total-signal controls provide the necessary parameters for the calculation of Z', a useful statistical parameter to assess the robustness of an assay *(13)*. A Z' value ≥ 0.5 is typically considered the acceptance limit for assay robustness for HTS *(10)*. However, there are assays, particularly cellular and complex secondary assays, where $Z' \geq 0.5$ may be an insufficient criteria for assessing assay performance. Gribbon et al. have described multiple aspects of quality control of screening data: intrawell, intraplate, and systematic trends in data analysis *(14)*. Systematic approaches to defining assay performance must be defined and applied postscreen to fully capture high-quality HTS data and form relevant conclusions *(15, 16)*.

The composition of a test sample is a key factor for HTS design and data analysis. Although HTS is commonly referred to in the context of screening large collections of small molecules to identify drug candidates, HTS concepts and technologies are also applied to screening of peptide, protein, RNA, or DNA samples. For example, hybridoma and monoclonal antibody screening is a powerful combination used to identify antibodies either as tools or as potential therapeutics *(17)*. Advances in molecular-biology techniques, coupled with chip technologies, have led to HTS applications of gene profiling to characterize up- or downregulation of genes in various cellular and disease states *(18)*.

Points to consider for small-molecule screening prior to initiating HTS and data analysis are the properties of the small-molecule library. Key factors for small molecules are (a) storage, (b) stability, and (c) structure. These points are beyond the scope of this chapter and are reviewed in the literature *(19–21)*. The challenge and costs of maintaining a large high-quality library have encouraged organizations to use smaller focused libraries, designed by molecular modeling, for testing against specific targets *(22, 23)*. Large compound collections remain in frequent use, as there is a wide range of molecular targets available for drug screening. The large library collections reflect greater diversity of chemical space. However, libraries fraught with degraded compounds increase the difficulty of data analysis.

Implementation of an assay technology platform is a key part of generating robust HTS data for analysis. HTS laboratories have many choices for establishing their infrastructure. Fully automated sample retrieval and assay screening approaches are available where liquid handling is integrated with the assay readout *(24, 25)*. Alternatively, stand-alone technology platforms can be utilized for liquid handling and assay readout. Based on the sample number and the difficulty of the assay, HTS may remain a largely manual process. This is often the case for secondary screens utilizing low-throughput assays to characterize the biological activity of compounds identified by HTS.

Points to consider for small-molecule screening prior to initiating HTS and data analysis are potential erroneous results due to the technology platform being utilized. For example, liquid-handling and storage needs differ significantly between 96- and 1,536-well plates *(26)*. Chip-based technologies and microfluidics have enabled new formats for biochemical and cellular assays *(27)*. HTS technologies must be matched carefully with the appropriate plate and assay format to ensure that datasets are free of technology-based artifacts.

The choice of which detection technology to use is often driven by the biological assay endpoint, coupled with the availability of assay reagents. Radioactive-based detection technologies continue to retain a place in HTS, due to the ability to directly label protein, enabling detection with minimal modification to a protein and thus enabling assays when alternate detection reagents are unavailable. However, in recent years advances in fluorescence detection have led to an increase in the availability of reagents for fluorescence-based biochemical, cellular, and whole animal endpoints *(28, 29)*. The introduction of new fluorescent tags that are coupled to key reagents enabled several new assay formats including time-resolved fluorescence resonance energy transfer and fluorescence polarization *(30, 31)*. These assay types primarily generated biochemical assays with high dynamic range, homogenous, and automation-friendly formats. Cell-based assays have also benefited from the increase in commercially available fluorescently tagged reagents. For example, high-content screening (HCS) assays utilize microscopic fluorescence cell imaging of multiple distinct fluorescent molecules *(32, 33)*. Though HCS may be low-throughput when compared with other assay technology platforms, the value lies with the use of multiple fluorophores, generating multiple endpoints. Cellular fluorescence imaging endpoints offer an attractive complementary assay format to commonly used cellular reporter-gene assays. Fluorescent technologies and their application in HTS and drug discovery have been reviewed extensively *(34, 35)*.

Data management and analysis are key components of extracting maximal value from an HTS dataset. Once the various components of HTS have been identified (compound collection, assay technology, assay parameters, etc.), one can determine the controls that will be used for the screening campaign. For example, if data is to be analyzed as % inhibition or % activation, control wells must be included that define total signal (100%) and background signal (0%). Quality control of HTS data can be based on the plate map and the assay controls used. It is vital that data be captured in a centralized database with criteria established for data entry. This facilitates analysis of datasets that are free of erroneous results and artifacts.

There are multiple approaches to identifying active compounds, or *hits*, for follow-up study from HTS *(36–40)*. Data analysis can be a largely automated or a manual process, depending on the software systems utilized. There are two frequently used approaches for identifying active compounds from HTS. First, active compounds can be defined as those that meet a given statistical criterion of activity or inhibition relative to a vehicle-control distribution. For example, if the assay endpoint is activation of a signaling event in a cell-based assay, three standard deviations from the mean of vehicle-control wells (designed to measure total signal in the absence of activating compound) can be used as the criterion for identifying active compounds. Since there will be well-to-well variations within a plate, the criterion might be expanded to include three standard deviations from the mean of a total-signal distribution including sample wells. This approach can be applied for generic library screens with success; however, when applied to focused libraries, in which sample wells generate a high hit rate, the approach can have limited value. Second, active compounds can be defined as those that meet an arbitrary readout criteria outside of the variability found within a given assay. For example, if the assay endpoint is the inhibition of cell proliferation at a given time point, an arbitrary definition of actives may be defined at those that have a % inhibition value of ≥50% inhibition. This approach can be beneficial when screening focused library collections or intentionally targeting a hit rate to accommodate follow-up studies.

Sophisticated statistical approaches based on computer-generated models of compound collections and activity can be utilized *(41)*. Relying solely on Z' and control wells for data analysis is a limited approach and does not provide the user with information about trends across multiple screens or compound collections. However, as control wells are the starting points for analysis of HTS datasets, this chapter focuses on the practical points of defining control wells and the plate maps for data analysis. These parameters can be readily expanded upon by more sophisticated statistical and model-driven approaches.

There are numerous options for plate maps, which can be utilized. Key determinants are the plate type, assay format, and availability of controls. Shown in **Fig. 1** are commonly used plate maps for HTS in a 96-well format.The map can be readily scaled up or down to accommodate alternate density of wells on a plate. The plate map for primary screening is designed to accommodate control wells and a single dose treatment of compound wells (**Fig. 1a**). Based on the availably of reagents for the primary screen and the assay robustness, one can determine the number of replicates needed to test sample wells. Replicates wells of $n= 2$ are recommended for primary HTS when assay reagents are available. The plate map is simply replicated in sample preparation to generate two daughter plates for HTS to provide the $n= 2$. The plate map for secondary screening is designed to accommodate control wells and multiple doses of compound for the generation of dose response curves (IC_{50} or EC_{50}) (**Fig. 1b**). The map shown allows for

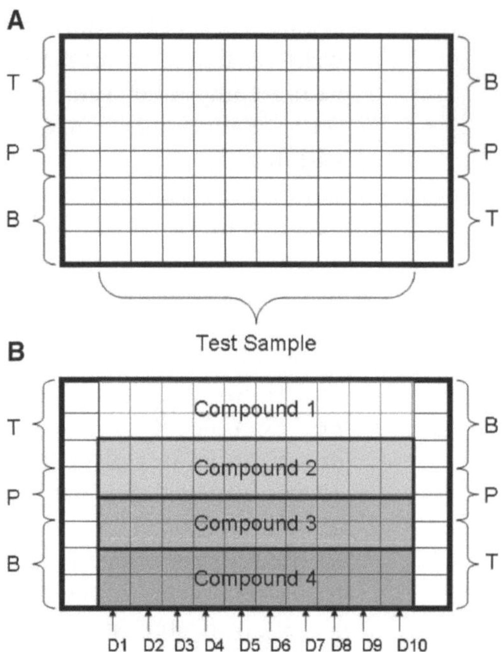

Fig. 1. Test compound sample and dose-response plate maps.

testing of four compounds in a ten-point dose response curve, where each dose is tested in duplicate. A ten-point dose curve can readily accommodate a 3–4 log concentration range of test compound, facilitating a full curve in a single measurement *(42)*. This plate map avoids multiple repeats to identify the appropriate dosing range of a given compound and provides sufficient data points to generate an acceptable curve fit. For example, a ten-point curve using a threefold dilution factor covers from 10 μM to ~0.5 nM (four logs). If throughput is a determining factor to the user for secondary screens, the plate map can be modified to accommodate fewer doses. For example, throughput will be doubled by modifying to a five-point dose curve. However, users must reconcile the quality of curve fit with the throughput obtained.

Control wells in the representative plate maps (**Fig. 1**) contain total-signal, positive-control, and background-signal wells. Total-signal wells and background-signal wells are key control wells, as they define the assay window and set the parameters for data quality control. Data from these control wells are critical in measuring assay robustness (i.e., Z') for a given plate. Thus, the majority of control wells on the plate map are dedicated to total signal ($n= 6$) and background signal ($n= 6$). Controls are divided into outer columns and inverted. This allows the user to review edge effects and signal drifts, and compensate accordingly. Positive-control wells are a desirable optional control, if available. These wells provide additional information regarding the assay window and can be particularly informative with assays that have a limited dynamic range. For example, for a given cellular assay measuring activation of a signaling event at a given time point, the positive control well can be defined as the 50% activation point measured by the assay. This layout provides the user with clear guidelines for the midpoint of the assay window on a given plate. If positive controls are unavailable (i.e., limited access to agonist/antagonist, or EC_{50} value is unknown), the plate map can be modified to eliminate positive-control wells in favor of additional total-signal and background-signal control wells.

1.2. Summary

A sample procedure for generating control wells and data analysis from a specified plate map for a 96-well SM HTS assay is described. This protocol utilizes standard liquid-handling approaches and can be readily modified to accommodate alternate density plate formats and screening detection technologies. Data analysis described is based on control wells typically utilized in HTS and provides a framework for further analysis with more sophisticated statistical and modeling approaches.

2. Materials

2.1. Library

1. Master storage plates with lyophilized test SM sample (*see* **Note 1**).
2. DMSO (Sigma-Aldrich, St. Louis, MO).

2.2. Assay and Automation

1. Biochemical or cellular assay for HTS (*see* **Note 2**).

2.3. Data Analysis

1. Database for capture of HTS data (*see* **Note 3**).
2. Software for data analysis (*see* **Note 4**).

3. Methods

3.1. Compound Library Preparation

3.1.1. Preparation of a Master Compound Plate

1. Resuspend powder of test compounds in 100% DMSO at 10 mg/mL or 10 mM in 96-well deep well plates or desired plate format. Fill wells based on test compound location of plate map (**Fig. 1a**). Label plates and track compound identification, well location, and plate label in database (*see* **Note 5**).
2. Mix to resuspend compounds by pipetting up and down. Use a single pipette tip for a single compound to avoid contamination of compounds from well to well.
3. Make a note in database of compounds that do not go into solution. This will serve as a flag for potential problems in assay data interpretation due to compound precipitation.
4. Store plates frozen at 4°C until ready for use. Stock plates can be stored for the duration of a screening campaign and utilized for preparation of plates for secondary screens. DMSO stocks will freeze at 4°C (*see* **Note 6**).

3.1.2. Preparation of Working Compound Plate

1. Prepare control-well solutions and add to specified wells of the plate map (**Fig. 1**). Identify liquid volumes based on HTS assay protocol. For total-signal control wells, options include (a) DMSO, (b) assay buffer, or (c) cell culture media. For background-signal control wells, options include (a) total signal including known inhibitor at specific concentration, or (b) assay reagents excluding a key reagent (cofactor, growth factor, etc.) needed to facilitate the reaction.
2. Prepare test compounds for single-dose (**Fig. 1a**) or dose-response (**Fig. 1b**) testing based on the plate map. Remove the master compound plate from storage and thaw to room temperature. When possible, use freshly made master plates. Make note in database of any new compound precipitation observed.

For HTS, screening concentrations typically vary from 1 to 50 μM. The dose is set higher for testing random large library collections, and lower for screening focused library collections. For IC_{50} or EC_{50} testing, a range of ten doses over 3–4 log units is typically used to generate a reliable curve. The upper limit is generally 10–50 μM. Higher doses of test compounds are subject to solubility problems and interference with assay-detection technologies. Introduce a mixing step for each serial dilution for the dose curve. For HTS or dose-response curve plate preparation, identify stock liquid and dilution buffer volumes, based on the assay protocol, and pipette into the working plate based on the plate map.

3. Store plates frozen at –20°C as needed. If possible, use working compound plates on the day of preparation.

3.1.3. Preparation of Compound Assay Plate

1. For assay protocols in which reagents are added directly to test compounds, prepare two mother–daughter compound plates for testing based on the assay protocol (*see* **Note 7**). Label plates in database for tracking purposes. Use plates immediately for screening (*see* **Note 8**).
2. For assay protocols in which compounds are added directly to an assay plate, draw replicate compounds for two mother–daughter assay plates directly from the working plate (*see* **Note 9**). Label assay plates in database for tracking purposes. In this scenario, addition of compound initiates the assay measurement, or an additional reagent step is needed to initiate the assay.

3.2. Assay

1. Run assay plates, keeping within the parameters of the HTS protocol. Examples of key assay parameters are light, time, temperature, and CO_2.
2. Recommendation: Include a control plate at the beginning and/or end of a given run (**Fig. 2**). Generate control plates based on total- and background-signal control wells (**Fig. 1**) being used for screening (*see* **Note 10**). Inclusion of a control plate allows the user to look for trends in a given testing run, such as a drift in the control-well signal.
3. Upon conclusion of assay, acquire assay readout based on endpoint being measured.

3.2.1. Database

Capture the following information (minimal requirements) in centralized database:

1. Assay name.
2. Assay description (*see* **Note 11**).
3. Screening dose.
4. Date.

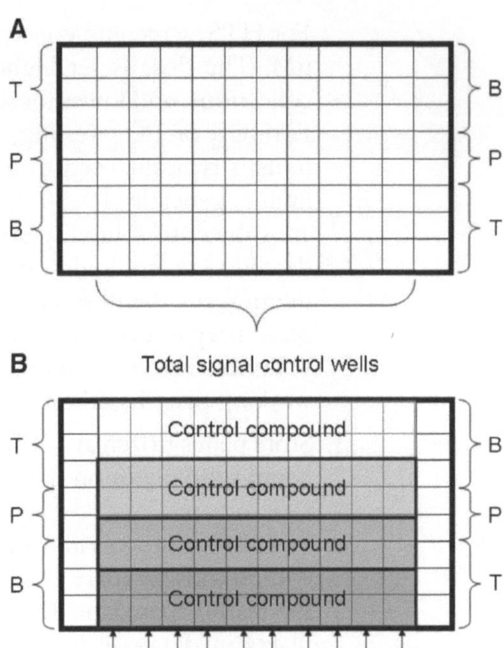

Fig. 2. Control compound sample and dose-response plate maps for assay development, or to accompany test compound sample plates in an HTS run.

5. User.
6. Plate format.
7. Plate map.
8. Compound source identification (working stock and master stock).
9. Data.
   a. Compound identification code.
   b. Screening raw data.
10. Key observations during screen (*see* **Note 12**).

3.2.2. Data Analysis

1. Calculate statistical parameters per plate using control wells on each plate.
2. Calculate signal/background (S/B) ratio and signal/noise (S/N) ratio:

$$S/B = \text{mean}X_2/X_1, \quad (1)$$

$$S/N = \text{mean of } X_2 - \text{mean}X_1/\text{standard deviation of } X_1. \quad (2)$$

where X_1 = minimal assay signal obtained from background control wells of plate map and X_2 = maximal assay signal obtained from total signal control wells of plate map.

3. Calculate Z' value:

$$Z' = 1 - (3S_2 + 3S_1)/X_2 - X_1. \quad (3)$$

where S_1 = standard deviation of X_1 and S_2 = standard deviation of X_2.

4. Using S/B, S/N, and Z' values, apply *pass* or *fail* criteria based on assay performance per plate. Identify plates ready for data analysis (*hit* identification or IC_{50} calculations). If data corresponding to control wells do not meet the *pass* criteria for a given plate, retest the compounds.

5. Calculate percent inhibition (% inhibition) or percent activation (% activity) for primary screen:

$$\% \text{ inhibition} = 100 \times (1 - (T - \text{mean}X_1)/(\text{mean}X_2 - \text{mean}X_1)), \quad (4)$$

$$\% \text{ activity} = 100 \times (T - \text{mean}X_1)/(\text{mean}X_2 - \text{mean}X_1)), \quad (5)$$

where X_1 = minimal assay signal obtained from background control wells of plate map, X_2 = maximal assay signal obtained from total signal control wells of plate map, and T = signal obtained from given compound test well of plate map. T can be based on $n = 1$ or as mean of n (i.e., $n = 2$ if duplicate measurements of test compound is obtained). The plate map and assay technology used will define the replicates necessary to establish T.

6. Identify active or *hit* compounds from the screen. Based on assay performance, identify threshold for active test compounds. Threshold for specific compound activity is defined per plate or per run based on the total-signal and background-signal control wells. Identify active compounds for follow-up characterization either by their ability to show an arbitrarily defined activity (i.e., 50% inhibition) or by their ability to show activity that is statistically significantly distinct from control wells. For example, define hit compounds as those where the test compound signal is three standard deviations above signal generated by background control wells (*see* **Note 13**).

7. Calculate IC_{50} or EC_{50} for dose-response activity in the secondary screen.

8. Plot fractional activity (y axis) of test compound as a function of the test compound concentration (X axis), based on the ten points of the plate map using curve-fitting software of choice.

9. Fit the data points to a curve using standard four-parameter logistic nonlinear regression analysis (i.e., GraphPad) (*see* **Note 14**).

10. Obtain IC_{50} or EC_{50} from curve fit; by definition, the test compound concentration that generates 50% inhibition of the total signal is the IC_{50} or EC_{50} value.
11. Ensure that there are sufficient data points defining the upper and lower plateaus of the curve to generate an accurate IC_{50} value. The advantage of the ten-point curve plate map is that there are multiple data points for the upper and lower plateau, and that it can define the slope of the curve. If there are not sufficient data points to define the curve, repeat the dose-response curve testing by shifting the dosing concentration as needed and/or modifying the fold dilution of the testing range to ensure full coverage by the dataset.

3.3. General Database and Analysis Management

1. Review data corresponding to compounds with solubility problems carefully. Insoluble compounds can form aggregates and generate spurious results in assays *(9)*.
2. A screening run with multiple plates may undergo a signal loss for total-signal control wells and a signal gain for background-signal control wells over the time course of the run. This effect reduces the assay window, and makes data analysis difficult. To prevent this scenario, perform runs within the range of assay performance (Z'). It may be necessary to run multiple small batches of assay plates for a given assay rather than long runs.

4. Notes

1. A variety of options are available from commercial suppliers of compound libraries for storage and handling. For handling of liquid DMSO solutions of test compound, use gloves and avoid exposure to test compounds. Compound collections may contain agents that are absorbed through skin via DMSO solution.
2. Use assays and automation with well-defined protocols. For example, for a biochemical assay establish parameters such as time point, temperature, and protein stability. For example, for a cellular assay, establish parameters particular to cell line, such as passage number, serum, and time points (with or without stimulation). Common assay-detection parameters to be established are detection reagent stability and stability of readout signal. This allows for flexibility in identifying the number of plates that can be accommodated in a given HTS run. Automation allows for greater precision and *hands-free operation*. However, for focused library collections and limited follow-up testing, automation may be limited to plate washers, multichannel pipettors, etc.

3. Typically, HTS campaigns capture data output in a single database linking chemical data to biological data, facilitating data analysis. Databases can either be generated in-house or purchased from commercial suppliers. It is beyond the scope of this chapter to define all database options. Commercial options include Accelrys® (www.accelrys.com) and IDBS (www.idbs.com). It is worth noting that many laboratories active in HTS generate a fusion between in-house and commercial databases.
4. Data-analysis tools are often integrated into databases (Accelrys Accord, IDBS ActivityBase). In addition, there are several stand-alone software packages, which are routinely applied. Spotfire is commonly utilized for assessing trends in HTS datasets. GraphPad (www.graphpad.com) and Kaleidagraph (www.synergy.com) are commonly utilized for secondary screens.
5. The efficiency with which test compounds are processed for HTS can be improved with a barcoding system. Speed and accuracy minimizes compound degradation and time spent in tracking compound plates during various stages of HTS.
6. If DMSO stocks do not freeze at 4°C, this likely is an indicator that the solution has incorporated water. The stock concentration is now likely inaccurate due to the presence of water. It can be difficult to determine the water content in the compound solution and thus recalculate an accurate compound concentration. Therefore, generate fresh DMSO stock solutions if sufficient test compound is available.
7. This configuration of reagent addition is more common for biochemical assays where proteins are directly added to DMSO or test compound. If multiple assay reagent additions are needed to initiate a reaction, a useful pipetting approach is to aspirate various reagents with an air gap separating each liquid followed by a single dispense. This approach can be used to add reagents simultaneously to test compound and initiate reaction.
8. Use freshly prepared plates to avoid loss of protein activity in the assay and/or compound degradation.
9. Ideally, the order of addition of assay reagents and compound will have been established during assay development for HTS. The addition of high concentrations of DMSO to proteins or cells can be a problem. Order of addition of assay reagents allows the user to control reaction kinetics ($t = 0$) and the introduction of DMSO to the assay.
10. Define the map for control plate (**Fig. 2**) using either total-signal wells (**Fig. 2a**) for a primary screen or control dose-response curve (**Fig. 2b**) for secondary screen. Utilize a plate map where the outer wells of the plate retain the control-well

configuration on each plate of an HTS run. Inclusion of a control plate allows user to identify and address edge effects, which can be pronounced during long testing runs. Corrective measures may be technical *(43)* or based on statistical data analysis *(8)*. For example, edge effects have been shown to be minimized in a cell-based assay by incubating freshly seeded plates of cells at room temperature prior to placing in an incubator.

11. A brief assay description with a link to a detailed protocol can be a valuable reference point to the user when comparing datasets from various assays. For example, if comparing cellular proliferation assay data across various cell lines, important factors for comparison are the cell type, readout (e.g., BrdU), time points, and starting cell densities. Ideally, these key parameters are captured in the database to facilitate queries across all assays.
12. Observations noted during the course of a given screen aid the user in performing data quality control of a given HTS run. Tracking events out of the ordinary (e.g., robotic crashes, clogging of liquid handlers, compound precipitation) allow the user to determine which data can be used and which compounds require retesting to acquire useful data.
13. Positive-control wells where the control compound is introduced to wells at its IC_{50} or EC_{50} value provide a benchmark for establishing the midpoint of the dynamic range of the assay on a given plate. Positive-control wells are a useful practical reference point for identifying a range of potency values for active compounds on a given plate; particularly if an arbitrary potency (i.e., 50% inhibition) is chosen as the criteria for the identification of active compounds. The contents of the positive-control wells can be modified readily to suit the potency range targeted by the user (e.g., 25%, 50%, 75%).
14. Shown below is a commonly used four-parameter logistic equation for curve fitting. Details can be found at http://www.graphpad.com and *(42)*.

$$Y = \text{Bottom} + \frac{(\text{Top} - \text{Bottom})}{1 + 10^{(\text{LogEC}_{50} - X)\text{HillSlope}}}.$$

References

1. Verkman, A. (2004) Drug discovery in academia. *Am. J. Physiol. Cell Physiol.* **286**, 465–474.
2. Root, D., Kelly, B., and Stockwell, B. (2002) Global analysis of large-scale chemical and biological experiments. *Curr. Opin. Drug. Discov. Dev.* **5**, 355–360.
3. Butcher, E. (2005) Can cell systems biology rescue drug discovery? *Nat. Rev. Drug Discov.* **4**, 461–467.
4. Liu, B., Li, S. and Hu, J. (2004) Technological advances in high-throughput screening. *Am. J. Pharmacogenomics* **4**, 263–276.
5. Archer, R. (2004) History, evolution and trends in compound management for high throughput screening. *Assay Drug Dev. Technol.* **2**, 675–681.
6. Echeverri, C. and Perrimon, N. (2006) High-throughput RNAi screening in cultured cells: A user's guide. *Nat. Rev. Genet.* **7**, 373–384.

7. Harding, D., Bradford, D., and Brown, G. (2002) Transferring industrial automation technology to the laboratory. *J. Assn. Lab. Automat.* 7, 84–88.
8. Malo, N., Hanley, J., Cerquozzi, S., Pelletier, J., and Nadon, R. (2006) Statistical practice in high throughput screening data analysis. *Nat. Biotechnol.* **24**, 167–175.
9. McGovern, S., Caselli, E., Grigorieff, N., and Shoichet, B. (2002) A common mechanism underlying promiscuous inhibitors from virtual and high throughput screening. *J. Med. Chem.* **45**, 1712–1722.
10. Carter, J. (ed.) (2003) A Guide to Assay Development. D&MD, Westborough, MA.
11. Rabitsch, K., Toth, A., Galova, A., Schleiffer, G., Schaffer, E., Aigner, C., Rupp, A., Penkner, A., Moreno-Borchart, A., and Primig, M. (2001) A screen for genes required for meiosis and spore formation based on whole-genome expression. *Curr. Biol.* **11**, 1001–1009.
12. Parng, C., Weng, W., Semino, C., and McGrath, P. (2002) Zebrafish: A preclinical model for drug screening. *Assay Drug Dev. Technol.* **1**, 41–48.
13. Zhan, J., Chung, T., and Oldenburg, K. (1999) A simple statistical parameter for use in evaluation and validation of high throughput screening assays. *J. Biomol. Screen.* **4**, 67–73.
14. Gribbon, P., Lyons, R., Laflin, P., Bradley, J., Chambers, C., Williams, B., Keighley, W., and Sewing, A. (2005) Evaluating real life high throughput screening data. *J. Biomol. Screen.* **10**, 99–107.
15. Sun, D., Whitty, A., Papadatos, J., Newman, M., Donnelly, J., Bowes, S., and Josiah, S. (2005) Adopting a practical statistical approach for evaluating assay agreement in drug discovery. *J. Biomol. Screen.* **8**, 508–516.
16. Eastwood, B., Farmen, M., Iversen, P., Craft, T., Smallwood, J., Garbison, K., Delapp, N., and Smith, G. (2006) The minimum significant ratio: A statistical parameter to characterize the reproducibility of potency estimates from concentration-response assays and estimation of replicate-experiment studies. *J. Biomol. Screen.* **11**, 253–261.
17. Lal, S., Christopherson, R., and Remedios, C. (2002) Antibody arrays: An embryonic but rapidly growing technology. *Drug Discov. Today* 7, 141–149.
18. Stegmaier, K., Ross, K., Colavito, S., O'Malley, S., Stockwell, B., and Golub, T. (2004) Gene expression-based high-throughput screening (GE-HTS) and application to leukemia differentiation. *Nat. Genet.* **36**, 257–263.
19. Archer, R. (2004) History, evolution and trends in compound management for high throughput screening. *Assay Drug Dev. Technol.* **2**, 675–681.
20. Chan, J. and Hueso-Rodriguez, J. (2002) Compound library management. *Methods Mol. Biol.* **190**, 117–127.
21. Holden, K. (2003) The significance of effective compound management. *Curr. Drug Discov.* **9**, 9–10.
22. Bajorath, J. (2002) Integration of virtual and high-throughput screening. *Nat. Rev. Drug Discov.* **1**, 882–894.
23. Sun, D., Chuaqui, C., Deng, Z., Bowes, S., Chin, D., Singh, J., Cullen, P., Hankins, G., Lee, W., Donnelly, J., Friedman, J., and Josiah, S. (2006) A kinase-focused compound collection: Compilation and screening strategy. *Chem. Biol. Drug Des.* **67**, 385–394.
24. Sundberg, S., Chow, A., Nikiforov, T., and Wada, H. (2000) Microchip-based systems for target validation and HTS. *Drug Discov. Today* **5**, 92–103.
25. Falconer, M., Smith, F., Surah-Narwal, S., Congrave, G., Lui, Z., Hayter, P., Ciaramella, G., Keighley, W., Haddock, P., Garethwaldron, G., and Sewing, A. (2002) High-throughput screening for ion channel modulators. *J. Biomol. Screen.* 7, 460–465.
26. Dunn, D., Orlowski, M., McCoy, P., Gastegeb, F., Appell, K., Ozgur, L., Webb, M., and Burbau, J. (2000) Ultra-high throughput screen of two-million member combinatorial compound collection in a miniaturized, 1536-well assay format. *J. Biomol. Screen.* **5**, 177–187.
27. Battersby, B. and Trau, M. (2002) Novel miniaturized systems in high-throughput screening. *Trends Biotech.* **20**, 167–173.
28. Bailey, S., Sabatini, D., and Stockwell, B. (2004) Microarrays of small molecules embedded in biodegradable polymers for use in mammalian cell-based screens. *Proc. Natl. Acad. Sci. USA* **101**, 16144–16149.
29. Miller, L. and Cornish, W. (2005) Selective chemical labeling of proteins in living cells. *Curr. Opin. Chem. Biol.* 9, 56–61.
30. Hertzberg, R. and Pope, A. (2000) High-throughput screening: A new technology for the 21st century. *Curr. Opin. Chem. Biol.* **4**, 445–451.
31. Ozkan, M. (2004) Quantum dots and other nanoparticles: What can they offer to drug discovery? *Drug Discov. Today* **9**, 1065–1071.
32. Bowen, W. and Wylie, P. (2006) Application of laser-scanning fluorescence microplate cytometry in high content screening. *Assay Drug Dev. Tech.* **4**, 209–221.
33. Dove, A. (1999) Drug screening-beyond the bottleneck. *Nat. Biotechnol.* **17**, 859–863.

34. Buttner, F. (2006) Cell-based assays for high-throughput screening. *Exp. Opin. Drug Discov.* **1**, 373–378.
35. Jager, S., Brand, L., and Eggeling, C. (**2003**) New fluorescence techniques for high-throughput drug discovery. *Curr. Pharm. Biotechnol.* **4**, 463–476.
36. Iversen, P. W., Eastwood, B. J., Sittampalam, G. S., and Cox, K. L. (2006) A comparison of assay performance measures in screening assays: Signal window, Z′ factor, and assay variability ratio. *J. Biomol. Screen.* **4**, 247–252.
37. Bleicher, K., Bohm, H., Muller, K. and Alanine, A. (2003) Hit and lead generation: Beyond high-throughput screening. *Nat. Rev. Drug Discov.* **2**, 369–379.
38. Walters, W. and Namchuk, M. (2003) Designing screens: How to make your hits a hit. *Nature Rev. Drug Discov.* **2**, 259–266.
39. Gagarin, A., Makarenkov, V., and Zentilli, P. (2006) Using clustering techniques to improve hit selection in high throughput screening *J. Biomol. Screen.* **11**, 903–914.
40. Woodward, P.W., Williams, C., Sewing, A., and Benson, N. (2006) Improving the design and analysis of high throughput screening technology comparison experiments using statistical modeling. *J. Biomol. Screen.* **11**, 5–12.
41. Konstantin, B. and Nikolav, S. (2006) Computational methods for analysis of high-throughput screening data. *Curr. Comp. Drug Des.* **2**, 1–19.
42. Copeland, R. A. (ed.) (2000) *Enzymes: A Practical Introduction to Structure, Mechanism and Data Analysis.* Wiley-VCH, New York.
43. Lundholt, B., Scudder, K. and Pagliaro, L. (2003) A simple technique for reducing edge effect on cell-based assays. *J. Biomol. Screen.* **8**, 566–570.

Chapter 14

Extracting Rich Information from Images

Anne E. Carpenter

Summary

Now that automated image-acquisition instruments (high-throughput microscopes) are commercially available and becoming more widespread, hundreds of thousands of cellular images are routinely generated in a matter of days. Each cellular image generated in a high-throughput screening experiment contains a tremendous amount of information; in fact, the name *high-content screening* (HCS) refers to the high information content inherently present in cell images (J Biomol Screen 2:249–259, 1997). Historically, most of this information is ignored and the visual information present in images for a particular sample is often reduced to a single numerical output per well, usually by calculating the mean per-cell measurement for a particular feature. Here, we provide a detailed protocol for the use of open-source cell image analysis software, *CellProfiler*, to measure hundreds of features of each individual cell, including the size and shape of each compartment or organelle, and the intensity and texture of each type of staining in each subcompartment. We use as an example publicly available images from a cytoplasm-to-nucleus translocation assay.

Key words: Complex phenotypes, Data visualization, High-content screening, High-throughput data analysis, Microscopy, Visual phenotypes.

1. Introduction

Each cellular image generated in a high-throughput screening experiment contains a tremendous amount of information. Historically, most of this information is ignored and the visual information present in images for a particular sample is often reduced to a single numerical output per well, usually by calculating the mean per-cell measurement for a particular feature. Researchers often want access to more than this data for a number of reasons:

P.A. Clemons et al. (eds.), *Cell-Based Assays for High-Throughput Screening, Methods in Molecular Biology, Vol. 486*

DOI: 10.1007/978-1-60327-545-3_1

- Even if a single measured feature is sufficient to score an assay, it may be unknown in advance *which* feature will be most useful for scoring the phenotype of interest, especially during the prototyping/assay development phase of a project.
- Even if a single measured feature is used to select hits, it is often useful to use other phenotypes measured from the same images as a virtual secondary screen, providing a richer characterization of each hit.
- The phenotype of interest might not be distinguishable using a single measured feature, but rather several per-image features inform the decision of whether to call a sample a hit. As an example, a minimum of N cells must be alive in the sample, in addition to the average staining brightness per cell being above a particular threshold. Obtaining several desired features can be challenging within some software packages. Frequently, prepackaged software designed for a particular assay requires that the cells be prepared with a particular set of stains, and is not flexible to certain kinds of multiplexing. In only some cases can this hurdle be worked around by combining the results of reprocessing the images with several separate software routines.
- The phenotype of interest might not be distinguishable using a single measured feature, but rather several *per-cell* features inform the decision of whether to call a sample a hit. As an example, a cell might need to be within a particular size range in addition to meeting a particular staining brightness threshold. This might especially be the case if the cell population being analyzed is mixed, and only cells of a particular type are to be used to make the primary measurement of interest. In some cases, tens to hundreds of features may be needed to properly classify cells as exhibiting a particular phenotype (for example, a complex morphology), using machine-learning techniques.
- A population-averaging technique other than mean might be required to distinguish samples of interest (e.g., median, percentage of cells above a threshold, Kolmogorov-Smirnov statistic).
- Thresholds for determining cells of interest might not be precisely known prior to the experiment. Most software processes images "on-the-fly," that is, concurrent with image acquisition. Thresholds determined in advance may not be suitable, in retrospect, for the entire experimental image set.
- Per-cell data itself might be of interest for downstream analysis, for example, for clustering or machine-learning techniques.
- Certain measured features may be useful to exclude artifactual samples or cells from contaminating an analysis.

In many cases, HCS software does not produce, or limits access to, the data necessary to score a particular phenotype, especially the per-cell data that is critical for more complex phenotypes. One alternative is for an expert biologist to score every image by eye, and indeed if the experiment is important enough or the visual phenotypes complex enough, this has been worthwhile *(2)*. Fortunately, automated image analysis has been improving. Many researchers have successfully used *CellProfiler*, a flexible open-source software package we have developed, to score meaningfully complex visual phenotypes for high-content screening. We have recently described and validated *CellProfiler* for many phenotypes in cell-based high-content screening assays *(3)* and for other biological samples *(4)*.

The basis for such screens is hundreds of measured cellular features for each individual cell in each individual image. The basic steps of an image analysis pipeline in *CellProfiler* are as follows:

1. Identify anomalies in the images, for example, uneven illumination of the sample or autofocus errors.
2. Correct those anomalies when possible.
3. Identify individual nuclei in each image (using images of a DNA stain – note that nuclei are typically more uniform in shape and more easily separated from one another than cells, so we first segment nuclei, then use segmented nuclei to seed the segmentation of individual cells).
4. Identify cell boundaries in each image (using images of a cellular stain, if available).
5. Identify subcellular organelles (using images of the organelle/structure stain).
6. Identify cellular subcompartments as desired (for example, cytoplasm and nuclear or plasma membrane).
7. Count the cells, measure the size and shape of each compartment or organelle, and measure the intensity and texture of each type of staining in each subcompartment. These measurements are exported to a database and are then available for downstream analysis using various computational tools.

Here, we explain the practical tips and techniques for applying *CellProfiler* to a cytoplasm-to-nucleus translocation assay. Many signaling pathways involve the translocation of protein from one subcellular compartment to another. In the assay used here, a fluorescently labeled *forkhead* protein (FKHR-EGFP) moves from the cytoplasm to the nucleus in response to positive-control drugs (1-h treatment with either wortmannin or LY294002) in human osteosarcoma cells, U2OS (**Fig. 1**). In this example, we show how to measure hundreds of features other than the primary readout (ratio of staining in the cytoplasm vs. nucleus) for downstream data analysis. This type of analysis revealed a change in nuclear

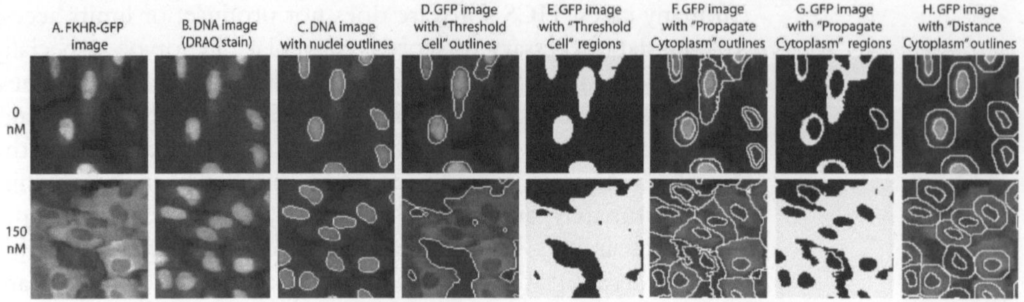

Fig. 1. Example images of the Cytoplasm to Nucleus Translocation assay. Top row = no drug; bottom row = 1-h incubation with 150 nM wortmannin. The actual images are approximately fivefold larger field of view in each dimension (640 pixels square), in 8-bit grayscale BMP format, with four replicates for each of the nine-point dose curve for each drug, plus negative controls. They are further described (*BioImage plate*) and were acquired from http://www.ravkin.net/SBS/CNT_Images.htm (courtesy of Ilya Ravkin). Outlines and regions are as processed by *CellProfiler* and saved using the OverlayOutlines or ConvertToImage module followed by a SaveImages module. Note that all images were contrast-enhanced using AutoLevels in Photoshop for display purposes only.

morphology in addition to the cytoplasm-to-nucleus translocation *(3)*. The steps involved in the protocol here are as follows:

1. Download and install *CellProfiler* software.
2. Download the example pipelines and run them on the example images.
3. Load the illumination correction pipeline, adjust it for your images, and run it.
4. Adjust the main pipeline for your images using test images.
5. Run the main pipeline on all images.
6. Export data.
7. Perform downstream data analysis.

It is much easier to adapt an existing pipeline than to build a new one from scratch, so we suggest using this cytoplasm-to-nucleus translocation pipeline as the basis for any HCS screen. In fact, it can be adapted to most cell types and assays using the point-and-click interface of *CellProfiler*. In some sense, the example pipeline we describe here is roughly similar no matter what the assay/phenotype of interest. The basic concept of identifying cells and measuring everything possible about them is constant from experiment to experiment; the uniqueness of each assay comes in determining which measure(s) are most useful to identify interesting samples, by data mining. A full discussion of appropriate data-mining methods is beyond the scope of this chapter, but we conclude the protocol by briefly reviewing the methods required to identify any particular phenotype of interest based on these measurements.

2. Materials

1. Fluorescence microscopy images to be processed. The setup time for an analysis is the same whether a handful or hundreds of thousands of images are processed. We routinely analyze tens of thousands of images per experiment. The images can be located within subfolders and need not be in a particular order or have a particular naming convention. A variety of file formats are currently readable by *CellProfiler*: *.bmp*, *.cur*, *.fts*, *.fits*, *.gif*, *.hdf*, *.ico*, *.jpg*, *.jpeg*, *.pbm*, *.pcx*, *.pgm*, *.png*, *.pnm*, *.ppm*, *.ras*, *.tif*, *.tiff*, *.xwd*, *.dib*, *.mat*, *.fig*, *.zvi*, *.flex*, *.stk*.
2. Computer (Macintosh, Windows, or Unix/Linux): A desktop or laptop computer with at least 1 GB of RAM (2 GB or more is preferable) and at least a 1-GHz processor is recommended.
3. *CellProfiler* software: The software can be downloaded for free (*see* **Subheading 3.1**).
4. Example images and their corresponding *CellProfiler* pipelines. These can be downloaded for free (*see* **Subheading 3.2**).
5. Computing cluster: This is optional, depending on the number of images to be processed and the processing speed for a particular analysis pipeline. The example image pipeline demonstrated here on 96 sample images requires about an hour on a single computer with a 2.6-GHz processor and 8-GB RAM. Larger image sets will likely require a computing cluster.

3. Methods

3.1. Download and Install CellProfiler Software

1. Choose whether to use the regular or developer's version.
2. Most users will use the regular version (also known as the "compiled"/binary version) suitable for their computing platform (Windows, MacIntosh, or Unix/Linux). This version is free (GPL license) and does not require a MATLAB license.
3. Researchers wishing to implement their own image-analysis algorithms should download the developer's version (that is, the MATLAB source code). The developer's version is free and open-source (GPL license), but does require installing MATLAB software and its license, not detailed here (Mathworks, Natick, MA; www.mathworks.com).

4. Download the version of software you chose from www.cellprofiler.org/download.htm.
5. Follow the installation instructions from the webpage to install *CellProfiler* and start it (*see* **Note 1**). If you encounter difficulties on this step, consult the installation instructions (www.cellprofiler.org/install.htm), or visit the online forum (www.cellprofiler.org/forum) to see if this problem has been encountered and solved before.

3.2. Download the Example Pipelines and Run Them on the Example Images

1. Download the example images and pipeline called *Human cytoplasm-nucleus translocation assay (SBS)* from www.cellprofiler.org/examples.htm. After downloading the file, make sure that it is *unzipped* (that is, uncompressed). Often, this occurs automatically, but if the file you downloaded does not automatically produce an accompanying folder, you should unzip the file manually. Usually, double-clicking the file accomplishes this (using your computer's built-in unzipping functionality); otherwise, use unzipping software. This will produce a folder containing the images described in Fig. 1.
2. Run the example pipeline called ExampleSBSPIPE.mat on the example images you downloaded. To do this, follow the instructions in Help > Getting Started > GettingStarted. This step allows you to see how processing typically proceeds (*see* **Notes 2–4**). If you obtain *Out of memory* errors, *see* **Note 5**.

3.3. Load the Illumination Correction Pipeline, Adjust It for Your Images, and Run It (see Notes 6 and 7)

1. After loading the ExampleSBS-IlluminationPIPE.mat pipeline into *CellProfiler*, make LoadImages module adjustments: Images need not be named or organized a particular way in order to use *CellProfiler*. Setting this module tells *CellProfiler* where to retrieve images and gives each image a meaningful name for the other modules to access. There are two basic methods to tell *CellProfiler* which batches of images are to be analyzed. The Order option is used when images are present in a folder or series of subfolders in repeating order (e.g., DAPI, GFP, DAPI, GFP, …) and the Text option is used when images have a particular piece of text in the name (e.g., all the DNA channel images contain the text "_D"; all the GFP channel images contain the text "_G"). Placing two different LoadImages modules in the pipeline allows you to choose one entire folder of a particular image type and a separate folder with a different image type (if GFP and DAPI images are stored in separate folders). There are a number of ways images can be loaded and identified; see the help for this module for more details (*see* **Note 8**).
2. CorrectIlluminationCalculate module adjustments: The only adjustment might be the two settings related to *width of artifacts*. The overall goal of this module is to produce an image that is

a believable representation of the illumination pattern that is affecting your images. The help for this module describes the various available methods, but Median Filtering is the most common method for HCS image sets. You will note that the example pipeline is set up to override the automatic calculation of artifact width, with a filter size of 150. If this image is too smooth (that is, it is not capturing the full amount of variation you think is present in the illumination pattern), then the filter size (or artifact width) should be decreased. If the image is too rough (that is, there is too much fine detail in the pattern such that it represents the random distribution of cells in your images rather than the real illumination pattern), then the filter size (or artifact width) should be increased, or you have an image set too small to accurately calculate the illumination pattern. If you have more than two channels for your image set, you should add additional CorrectIlluminationCalculate and SaveImages modules to the pipeline.

3. SaveImages module adjustments: no adjustments should be needed.
4. Once the modules are appropriately set, run the adjusted illumination correction pipeline to produce illumination correction images for your image set. Make sure that the default image and output folders are appropriately set, and click Analyze images. After processing completes, the resulting illumination correction images (Channel1ILLUM.mat and Channel2ILLUM.mat) should be produced and deposited in the default output folder (*see* **Notes 9** and **10**).

3.4 Adjust the Main Pipeline for Your Images Using Test Images (see Notes 11 and 12)

1. Using your computer's normal interface, create a test folder and copy several test images into it (*see* **Note 13**).
2. Using your computer's normal interface, create a test output folder.
3. In *CellProfiler*, set the default image and output folder to be your test image folder and test output folder, respectively (*see* **Note 14**).
4. LoadImages module adjustments: *see* **Subheading 3.3**, **step 1**.
5. LoadSingleImage module adjustments: make sure that the illumination-correction images you produced as described in **Subheading 3.3**, **step 4**, are specified to be loaded by this module.
6. LoadText module adjustments: first, you must produce an appropriate text file containing the doses corresponding to your images (*see* **Note 15**). This is necessary for the CalculateStatistics module to tell you how effective the assay is overall. Using the example file 1049_Dose.txt as a template (*see* **Note 16**), adjust it so that each line of the text file corresponds to the dose for each image cycle (each field

of view) that you are processing, in the order that the cycles will be processed. Looking in the window in the bottom left on the main *CellProfiler* window shows you the order that your images will be processed. If you have only positive and negative controls rather than a full dose curve, you can use a dose of 1 for positives and 0 for negatives in the text file. Once you have created the text file and placed it in the project folder, be sure that the LoadText file is set to use the text file you just made (*see* **Note 17**).

7. IdentifyPrimAutomatic module adjustments: there are two IdentifyPrimAutomatic modules in this pipeline – the first identifies nuclei using the DRAQ channel (**Fig. 1C**) and the second simply identifies the portions of the cells that show GFP staining (**Fig. 1D** and **E**). It turns out that for the primary readout of the assay (ratio of GFP staining in cytoplasm vs. nucleus), the second is not useful, but we include it here under the principle of measuring a large number of features of each image and choosing later what is useful. Several methods for object identification are available in a modular fashion within the IdentifyPrimAutomatic modules (*see* **Note 18**). The first IdentifyPrimAutomatic module is probably the most important to adjust in getting the pipeline to work well on your image set, so take time to see the help for the module for a complete description of the available methods and tips for the settings. In general, the settings that might require adjustment will be as follows:
   (a) The *typical diameter of objects*; the *automatic thresholding method*, *threshold correction factor*, and *approximate percentage covered by objects* (if *CellProfiler* appears to be inaccurately deciding between nuclei and background).
   (b) The *lower and upper bounds on threshold* (very important to be set properly to prevent poor results on images that are extremely different from expected, usually images with dramatic artifacts).
   (c) The *size of smoothing filter* and *suppress local maxima* (if *CellProfiler* is inappropriately splitting nuclei in two or merging clumps of cells together).
8. IdentifySecondary module adjustments: there are two IdentifySecondary modules in this pipeline – the first finds the edges of the GFP staining (**Fig. 1F** and **G**). In samples where GFP is primarily nuclear, the *objects* found by this module will simply be the nuclei, but in samples where the GFP staining is primarily cytoplasmic, the edges of the *objects* will extend to cell edges. The second IdentifySecondary module creates a ring around each nucleus, which is used as a proxy for cell edges in the absence of a cellular stain (**Fig. 1H**; *see* **Note 19**). In general, the settings that might require adjustment for the

first IdentifySecondary module (Propagate option) will be as follows:

(a) The *automatic thresholding method*, *threshold correction factor*, and *approximate percentage covered by objects* (if *CellProfiler* appears to be inaccurately deciding between cells and background).

(b) The *lower and upper bounds on threshold* (very important to be set properly to prevent poor results on images that are extremely different from expected, usually images with dramatic artifacts).

(c) The *regularization factor* (this controls the precise dividing line between cells that touch each other). In general, the only setting that might require adjustment for the second IdentifySecondary module (Distance option) will be *number of pixels to expand* (this controls the size of the rings around each nucleus).

9. IdentifyTertiarySubregion module adjustments: no adjustments should be necessary.
10. MeasureCorrelation module adjustments: no adjustments should be necessary, unless channels are available beyond DNA and GFP, or unless new compartments have been defined.
11. MeasureObjectIntensity module adjustments: no adjustments should be necessary, unless channels are available beyond DNA and GFP, or unless new compartments have been defined.
12. MeasureObjectAreaShape module adjustments: no adjustments should be necessary, unless new compartments have been defined.
13. MeasureTexture module adjustments: this module should be adjusted if channels are available beyond DNA and GFP, or if new compartments have been defined. Further, the scale of the texture (in units of pixels) should be adjusted for a particular image set. The scale of the texture is essentially empirical; a higher number measures larger patterns of texture, whereas smaller numbers measure more localized patterns of texture (fine texture). You are essentially asking *CellProfiler* to tell you whether the staining pattern is smooth on a particular scale. You can also add several texture modules, each measuring a different scale of texture, to the pipeline. See the help for this module for more details.
14. CalculateRatios module adjustments: no adjustments should be necessary, unless channels are available beyond DNA and GFP, or unless new compartments have been defined.

15. ClassifyObjects module adjustments: the threshold needs to be set empirically for this module. Of course, as has been stated as the overall approach here, we recommend choosing thresholds postprocessing, using appropriate data exploration/analysis tools, but if the threshold is already close to known, it can be useful to classify cells as positive or negative within the pipeline, so that CalculateStatistics can be used on the resulting data.
16. CalculateStatistics module adjustments: *see* **Subheading 3.4, step 6**.

3.5. Run the Main Pipeline on All Images

1. If the number of images is manageable for a single computer, change the image folder from the test images folder to the real images folder; change the output file name (if desired), and click the Analyze images button.
2. If the processing time would be too great on a single computer, then run the processing on a cluster as follows:
   (a) Download and install *CellProfiler* on a computing cluster. See the installation instructions as well as Help > General Help > Batch Processing within *CellProfiler* and the online forum (www.cellprofiler.org/forum). The developer's version may be used on an entire cluster if MATLAB licenses are available. Alternatively, you should use a compiled version. There are a wide variety of computing clusters in existence; one compiled version of *CellProfiler* for cluster computers is available for download at www.cellprofiler.org. If this version is not compatible with a particular cluster, the developer's version (source code) can be downloaded and recompiled (using MATLAB's compiler) on a representative cluster computer. This will require a single MATLAB license, but the resulting compiled code can be run on the entire cluster without MATLAB licenses.
   (b) Add the ExportToDatabase module (optional). When splitting your images into batches, your data will be produced in separate data files for each batch and you have two options:
      - If the resulting data files are not overwhelmingly large, you can merge the output files into a single output file (*see* **Subheading 3.6, step 2**). This option does not require adding any additional modules to the pipeline.
      - Most often for large image sets, you will prefer to export the resulting data to a MySQL or Oracle database for further analysis and exploration. In this case, add the ExportToDatabase module to the pipeline and configure it according to the help for the module.

(c) Add the CreateBatchFiles module to the end of the pipeline and configure it appropriately, according to the help for the module.

(d) Run the first image cycle locally. Clicking the Analyze images button will cause *CellProfiler* to process the first batch of images locally, and then produce the necessary files for batch processing.

(e) Submit the batches to your cluster for processing. See the Help > General Help > Batch Processing within *CellProfiler* for details.

(f) Check the progress of processing. For example, if using LSF software to submit jobs to your cluster, *see* **Table 1** for some basic commands to monitor the progress of processing.

3.6. Export Data

Once processing has completed, the data can be exported in three ways:

1. If you ran all images on one computer, the data can be exported to Excel using the data tool in the main window (Data Tools > ExportData) (*see* **Note 20**).

Table 1
Commands to monitor processing of jobs when using LSF software for cluster computing

<table>
<tr><th>Goal</th><th>Command</th></tr>
<tr><td>List all jobs</td><td>bjobs</td></tr>
<tr><td>Count all jobs</td><td>bjobs | wc -l</td></tr>
<tr><td>Count running jobs</td><td>bjobs | grep RUN | wc -l; echo jobs_running</td></tr>
<tr><td>Count pending jobs</td><td>bjobs | grep PEND | wc -l; echo jobs_pending</td></tr>
<tr><td>Check how many jobs are running and pending for a particular sample</td><td>bjobs -w | grep samplename | wc -l</td></tr>
<tr><td>Kill all jobs</td><td>bkill 0</td></tr>
<tr><td>Switch job to another queue</td><td>bswitch normal 189091 (where 189091 is the job number and normal is the name of the desired queue to switch to)</td></tr>
<tr><td>Check how many jobs are running and pending in a particular queue</td><td>echo normal; bjobs | grep normal | wc -l (where normal is the name of the queue of interest)</td></tr>
<tr><td>See the output of a job while it is still running</td><td>bpeek</td></tr>
<tr><td>To submit an individual job</td><td>bsub -B -N -u email@X.edu matlab -nodisplay -r Batch (Notes: -B sends email at beginning of the job, -N at the end; -q QUEUENAME allows selecting a queue; BATCH is the name of the job, for example Test3Batch_2_to_2.)</td></tr>
</table>

2. If you ran the batches on a cluster, but the sizes of the resulting output files are not overly large, use DataTools > MergeOutputFiles to merge the output files into a single file, and then use DataTools > ExportData to export the data to a tab-delimited file that you can open in Microsoft Excel (*see* **Note 20**).
3. If you ran batches on a cluster and used the ExportToDatabase module, the data can be exported by following the instructions for the ExportToDatabase module.

3.7. Perform Downstream Data Analysis

As mentioned in the introduction, we briefly overview options here:

1. Challenges of image-based data analysis: Typical screening experiments include tens to hundreds of thousands of images and each image contains hundreds or thousands of cells. *CellProfiler* measures hundreds of features about each of these millions of cells. Drawing meaningful conclusions from this data beyond very well-defined, simple analyses can be very challenging, due both to the huge amount of data involved and to the high dimensionality of the measurements. Furthermore, biases in the data can occur due to variation across the physical layout of the multiwell plates (known as plate or edge effects). Bias correction is an active area of research *(5–7)* and beyond the scope of this chapter.
2. Types of image-based data analysis: The introduction lists some of the more complex types of data analysis that might be necessary to identify samples of interest. Here, we group and describe some of these types of analyses. Note that data exploration itself is often a necessary step in the process of choosing an analysis method, and after hits are chosen in the screen, further data exploration may be of interest to characterize hits.
   a. Per-image averaging: Per-cell measurements can be combined to give per-image values by taking means, medians, etc., or otherwise reducing each measurement to a small set of parameters. This approach can work well for highly penetrant phenotypes. Several per-image features may be necessary to identify samples of interest.
   b. Per-image population comparisons: Each sample's cell population can be compared to a control cell population using distribution-based metrics such as Kolmogorov-Smirnov *(8)* or Kuiper *(9)* tests, or compute sample-based information-theoretic estimates, such as the KL divergence between the two distributions *(10)*.
   c. Per-cell classification: Individual cells can be classified as fitting a particular phenotype, based on the cell's measurements. The proportion of cells in each sample fitting the phenotype is thus the primary readout. This type of analysis is able to detect very small changes in the per-

centage of cells showing a particular phenotype (for example, from 1 to 3%), which is highly relevant considering that cell populations are inherently highly variable *(11)*. Moreover, reducing per-cell measurements to per-image measurements ignores the relationships between measured features on a per-cell basis. These relationships can sometimes be important for distinguishing samples of interest. For example, two images may have identical proportions of bright-staining and dark-staining cells in the green and red channels, but in one sample, the bright green cells tend also to be bright red, while in the other, bright green cells tend to be dark red. Such phenotypes would be impossible to distinguish using the per-image approaches described earlier. There are two options for per-cell classification:

- If a small number of measured features are sufficient to identify a cell subpopulation of interest, and if these measured features are known in advance, the subpopulation of interest can be identified in scatter plots of per-cell data, with the axes chosen by the researcher. Successive gating (choosing a subpopulation in a scatter plot, replotting that subpopulation on a new scatter plot with two new axes) can make use of more than just two measured features per cell.
- When more than just a few features are needed to define the cell subpopulation of interest, or if these features are not known in advance, machine-learning methods can be used to train a computer to recognize the phenotype on a per-cell basis. We have had success using an approach similar to content-based image retrieval in the field of computer vision *(12, 13)*. The biologist sorts cells as positive or negative for their phenotype of interest, and the software develops rules that are able to distinguish these examples based on the rich set of per-cell data collected by *CellProfiler*. The biologist sorts cells until the classifier becomes sufficiently accurate, and then all cells in all images are scored automatically by the computer as positive or negative. Images are then scored based on the number or percentage of phenotype-positive cells.

3. Tools for image-based data analysis: For data exploration and analysis from high-content screens, interactivity is needed between per-image data, per-cell data metadata for each sample, and raw images. The amount of data, (in terms of samples, cells, and features) pushes the limits of most software packages. We have recently released open-source data exploration software called *CellProfiler Analyst* (www.cellprofiler.org) readily enable the data analysis approaches

described above, especially that in **Subheading 15.3.7, step 2c**. At the time of this writing, other software is available for some of the analyses described above, including Spotfire (www.spotfire.com), CellMine (www.bioimagene.com), GeneData Screener (www.genedata.com), in addition to software sold in conjunction with high-throughput microscopes. Skilled researchers can also query databases directly to analyze data as needed.

4. Notes

1. If some of the text in the main window is not easy to see, go to File > Set Preferences to permanently set a more appropriate font size (*see* **Note 21**).
2. You will note that one display window opens for each module in the pipeline, and that as each field of view (*cycle*) is processed, the window refreshes to display the current cycle. The Status window remains open during processing and indicates how many cycles will be performed, how many are completed, and the amount of time elapsed. It also gives you the option to pause or cancel processing. The Details button indicates how much time is consumed on a per-module basis.
3. The measurement data acquired during each image cycle is stored at the end of every cycle in an output file, which grows larger and larger as the cycles complete. You can name the output file using the box next to *Analyze images* in the main window. The default output file name is DefaultOUT.mat, and it is impossible to overwrite an existing output file – if you attempt to do this, *CellProfiler* will ask if it is acceptable to add a number to the file name so that it is a new, unique file name.
4. Processing is much faster if display windows are closed. They may be closed during processing, but it is best to wait until the first cycle is finished to maintain the ability to reopen them properly later, using the Open figure and Close figure buttons in the main window. You can also go to File > Set Preferences to change the Display Mode if you would like to choose which windows to display each time you process an image set, or if you would like to *not* display windows, by default (*see* **Note 21**).
5. Image processing, especially for large image sets, is quite memory-intensive, and at some point during processing as measurements accumulate, you may run out of memory on your computer. If so, try the following:

(a) Close all other programs on the computer before starting *CellProfiler*.

(b) Run the example images on a small test set of images. You can remove a subset of the images from the folder and process just the remaining ones. You will also need to adjust the Dose text file accordingly – *see* **Subheading 3.4, step 6**.

(c) Use a different computer with more memory.

(d) For more tips, *see* Help > General Help > Memory and Speed within *CellProfiler*.

6. The physical limitations of any microscope lead to nonuniformity in the optical path of the sample, microscope, and/or its camera. We use the term *illumination anomalies* to refer to all sources of nonuniformity, including an uneven light source, uneven filters or lenses, and nonuniform camera sensing. Most software controlling the microscopes has the ability to perform a white-referencing step, where an image of a uniformly fluorescent (*flat*) sample is used to calculate and then correct for the illumination anomalies. This option is often ignored or improperly calibrated. Even when images are supposedly corrected by this method, these variations are frequently greater than 1.5-fold across the field of view, and can severely compromise the accuracy of measurements obtained. The standard illumination correction described here is based on the assumption that all the images in the experiment were collected in a single acquisition run, where as many elements as possible are constant; for example, the optical path and filters have not been moved, the lamp has maintained constant intensity, and the multiwell plate bottoms and the location imaged within each well are uniform. It is also based on the assumption that the background staining is negligible relative to the signal, and that the distribution of cells within each image is uniform (that is, there are no consistent biases of cells in one particular location). We have developed methods to detect and correct for nonuniform cell distributions but, while effective, they have not yet been made easily usable and are not discussed here (Thouis R. Jones, AEC, David M. Sabatini, and Polina Golland, unpublished data). For the standard illumination-correction calculation, we estimate the illumination variation as a smoothed per-channel average of all the images in the screen. See our other work for more details on the theoretical basis of this approach *(14)*.

7. To test whether the illumination stayed relatively constant across an experiment, you can calculate the illumination correction images as described here for subsets of images – a

set of images at the beginning and a separate set of images at the end of the experiment. If the overall pattern is relatively similar, then it is experimentally justified to apply the images calculated *from* the entire set *to* the entire set. If the patterns are substantially different from the beginning to the end of the experiment, you will need to develop a more sophisticated method of illumination correction, or make an attempt to find the point in time where the illumination pattern changed by calculating the illumination correction images from several subsets of images across the experiment. Beware, however, that this calculation will not work well for too small a subset of images. *See* **Subheading 3.3**, **step 2**, for a further description of how to tell whether an image set is *too small* for good calculation.

8. Color images can also be processed in *CellProfiler*. After loading the color images with the LoadImages module, use the ColorToGray module to split the images into their component colors.

9. When image sets are large, there are two ways to speed up the calculation of the illumination correction pipeline: you can run a separate pipeline for each wavelength on individual local computers, or on remote cluster computers. To use remote computers, add the CreateBatchFiles module to the end of the pipeline (*see* **Subheading 3.5**, **step 2**). In that module, set the number of images per batch to be larger than the total number of image cycles in your image set, because each wavelength's processing must be performed in a single batch job.

10. The display window for the CorrectIllumination_Calculate modules will show in red the minimum and maximum pixel intensity for the calculated illumination correction images. If the illumination varies more than 5- to 10-fold across the field of view (that is, the minimum is 1 and the maximum is more than 5–10), you should probably consider improving image acquisition to make the illumination more uniform, or cropping the images (using the Crop module, Rectangle option) to remove the worst parts of the images.

11. For help or a technical explanation for any module, click the "?" button in the main *CellProfiler* window below the pipeline, while the module of interest is selected, or go to Help > Modules Help and choose the module of interest.

12. There are multiple modules for measurement in the pipeline, even though not all of them may be useful to score a particular assay (see the introduction for the rationale). This pipeline measures the size and shape of each cell (e.g., area, perimeter, extent, convexity, and several Zernike moments)

and the intensity and texture of the various stains (e.g., mean and standard deviation of intensity, correlation of stains, and Gabor filter response at various scales). Measurements are made for each cellular compartment that has been defined (e.g., nucleus, cytoplasm, and entire cell).

13. Good images to use in your test set include negative and positive controls and images from the beginning and end of an acquisition run. In general, you want the very *worst*-case images to be included in your testing, not the *best* ones. The biggest pitfall in setting up cell image analysis is overtuning the analysis to a small, nonrepresentative set of images.
14. Browsing or typing into the main window will set the default image and output folders for your current session. To set them semipermanently, go to File > Set Preferences to select an image and output folder (*see* **Note 21**).
15. If the image set you are processing is not a dose-response experiment or a positive/negative control experiment, the LoadText module and CalculateStatistics module can be removed from the pipeline altogether.
16. The file included in the example image set, 1049_Dose.txt, contains the dose information for the example images. Note that the first half of the example image set is a dose-response curve for the drug wortmannin, and the second half is a dose-response curve for LY294002. For simplicity, the example pipeline described in this article processes both halves of the experiment together and therefore the CalculateStatistics module calculates the assay statistics across both drugs. For this reason, we have entered the doses in 1049_Dose.txt as 0–10 rather than their actual absolute concentrations in nM, to allow the entire image set to be run together, and to allow the calculation of a single Z factor for both positive controls. Note that, strictly speaking, the V factors and EC_{50} values from the example image set are not exactly interpretable, because we have combined the dose-response curves from two different drugs.
17. You can enter doses as exponents in the text file and then set the CalculateStatistics module to log-transform values. See the CalculateStatistics module for more details on how to set up the text file. In addition, for time-lapse image series, time points can be used as doses.
18. Very unusual cell types might require the development of new algorithms, which can be written and integrated into the open-source code of *CellProfiler*, given its modular and flexible structure. See Help > Developer Info in the main window of *CellProfiler*.

19. The ideal approach would be to collect images of a cellular stain (e.g., actin) and use IdentifySecondary's Propagate option to determine the edges of cells based on the cell stain image. The Propagate option is typically preferred over the Watershed method because Propagate does not rely on the borders of cells being consistently bright or dark, and it is less sensitive to gaps in staining *(15)*.
20. You can also put the module ExportToExcel in the pipeline to automatically export the data to Microsoft Excel files at the end of processing. Note that the files produced are tab-delimited, and can be opened by any spreadsheet program, not just Microsoft Excel. Warning: many versions of Microsoft Excel are limited to 256 columns and 65,536 rows, so your data may be too large to open in this program. Swapping rows and columns when exporting might be a good workaround if you reach these limits in only one dimension.
21. You can save a different set of preferences for each user, or even for each project, by choosing not to save the preferences as default, but rather giving them a specific name, *User1Preferences.mat*, for example.

Acknowledgments

The author is grateful to David M. Sabatini and his laboratory members at the Whitehead Institute for Biomedical Research, and Polina Golland and Thouis R. Jones at the Massachusetts Institute of Technology for their support and intellectual contributions to the methods described in this work. Images from Ilya Ravkin and testing by Adam Papallo are also much appreciated. The methods described in this work were also developed with the support of academic grants from the Society for Biomolecular Sciences and L'Oreal for Women in Science, and a Novartis fellowship from the Life Sciences Research Foundation.

References

1. Giuliano, K. A., DeBiasio, R. L., Dunlay, R. T., Gough, A., Volosky, J. M., Zock, J., et al. (1997) High-content screening: a new approach to easing key bottlenecks in the drug discovery process. *J. Biomol. Screen.* **2**, 249–259.
2. Kiger, A., Baum, B., Jones, S., Jones, M. R., Coulson, A., Echeverri, C., et al. (2003) A functional genomic analysis of cell morphology using RNA interference. *J. Biol.* **2**, 27.
3. Carpenter, A. E., Jones, T. R., Lamprecht, M. R., Clarke, C., Kang, I. H., Friman, O., et al.

(2006) CellProfiler: image analysis software for identifying and quantifying cell phenotypes. *Genome Biol.* 7, R100.

4. Lamprecht, M. R., Sabatini, D. M., and Carpenter, A. E. (2007) CellProfiler: free, versatile software for automated biological image analysis. *Biotechniques* **42**, 71–75.
5. Root, D. E., Kelley, B. P., and Stockwell, B. R. (2003) Detecting spatial patterns in biological array experiments. *J. Biomol. Screen.* **8**, 393–398.
6. Makarenkov, V., Kevorkov, D., Zentilli, P., Gagarin, A., Malo, N., and Nadon, R. (2006) HTS-Corrector: software for the statistical analysis and correction of experimental high-throughput screening data. *Bioinformatics* **22**, 1408–1409.
7. Malo, N., Hanley, J. A., Cerquozzi, S., Pelletier, J., and Nadon, R. (2006) Statistical practice in high-throughput screening data analysis. *Nat. Biotechnol.* **24**, 167–175.
8. Perlman, Z. E., Slack, M. D., Feng, Y., Mitchison, T. J., Wu, L. F., and Altschuler, S. J. (2004) Multidimensional drug profiling by automated microscopy. *Science* **306**, 1194–1198.
9. Kuiper, N. H. (1962) Tests concerning random points on a circle. *Proc. K. Ned. Akad. Wet. Ser. A* **63**, 38–47.
10. Kullback, S. and Leibler, R. A. (1951) On information and sufficiency. *Ann. Math. Stat.* **22**, 79–86.
11. Levsky, J. M. and Singer, R. H. (2003) Gene expression and the myth of the average cell. *Trends Cell Biol.* **13**, 4–6.
12. Muller, H., Michoux, N., Bandon, D., and Geissbuhler, A. (2004) A review of content-based image retrieval systems in medical applications – clinical benefits and future directions. *Int. J. Med. Inform.* **73**, 1–23.
13. Tieu, K. and Viola, P. (2004) Boosting image retrieval. *Int. J. Comput. Vis.* **56**, 17–36.
14. Jones, T. R., Carpenter, A. E., Sabatini, D. M., and Golland, P. (2006) Methods for high-content, high-throughput image-based cell screening, in *Proceedings of the Workshop on Microscopic Image Analysis with Applications in Biology (MIAAB)* (Metaxas, D. N., Rittcher, J., and Sebastian, T., eds.) Copenhagen, Denmark, pp. 65–72.
15. Jones, T. R., Carpenter, A. E., and Golland, P. (2005) Voronoi-based segmentation of cells on image manifolds, in Proceedings of the ICCV Workshop on Computer Vision for Biomedical Image Applications (CVBIA) (Liu, T. J., Zhang, C., eds.) Springer, Berlin, Germany, pp. 535–543.

Index

A

B

C

D

E

F

G

H

I

K

L

M

S

T

V

W

Y

Z

MIX
Papier aus verantwortungsvollen Quellen
Paper from responsible sources
FSC® C105338

If you have any concerns about our products, you can contact us on
ProductSafety@springernature.com

In case Publisher is established outside the EU, the EU authorized representative is:
Springer Nature Customer Service Center GmbH
Europaplatz 3, 69115 Heidelberg, Germany

Printed by Libri Plureos GmbH
in Hamburg, Germany